ORGANIC SPECTROSCOPY

ORGANIC SPECTROSCOPY

William Kemp

Senior Lecturer in Organic Chemistry
Heriot-Watt University, Edinburgh

A HALSTED PRESS BOOK

JOHN WILEY & SONS
New York

© William Kemp 1975

First published in the United Kingdom 1975 by
The Macmillan Press Ltd

Published in the U.S.A. by
Halsted Press, a Division of
John Wiley & Sons, Inc.,
New York

Printed in Great Britain

Library of Congress Cataloging in Publication Data

Kemp, William, 1932–
   Organic spectroscopy.
   "A Halsted Press book."
   Includes bibliographies and index.
   1. Spectrum analysis. 2. Chemistry, Organic.
I. Title.
QD272.S6K45 1975   547′.308′5   74–7675
ISBN 0 470–46842–4

TO LOUIE

CONTENTS

PREFACE

This book is an introduction to the application of spectroscopic techniques in organic chemistry. As an introduction it presupposes very little fore-knowledge in the reader and begins at a level suitable for the early student. Each chapter is largely self-contained, beginning with a basic presentation of the technique and developing later to a more rigorous treatment. A supplement to each of the principal chapters covers recent and recondite areas of the main fields, so that the book will also serve to refresh and update the postgraduate student's knowledge. Sufficient correlation data are given to satisfy the average industrial or academic user of organic spectroscopic techniques, and these tables and charts constitute a useful reference source for such material.

SI units are used throughout, including such temporarily unfamiliar expressions as *relative atomic mass* and *relative molecular mass*. A major break with the British conventions in organic nomenclature has also been made in favour of the American system (thus, l-butanol rather than butan-l-ol). This step is taken both in recognition of the vast amount of chemical literature that follows American rules (including the U.K.C.I.S. computer printouts) and in the expectation that these conventions will in due course be adopted for use by more and more British journals and books.

Chapter 1 takes a perspective look at the electromagnetic spectrum, and introduces the unifying relationship between energy and the main absorption techniques.

The next four chapters deal with the four mainstream spectroscopic methods—methods which together have completely altered the face of organic chemistry in little over a decade and a half. Most students will use infrared spectroscopy first (chapter 2), and nuclear magnetic resonance spectroscopy will follow (chapter 3). Electronic spectroscopy is more limited in scope (chapter 4), and mass spectroscopy (chapter 5) is the most recent and, in general, the most expensive. Chapter 6 provides both worked and problem examples in the application of these techniques, both singly and conjointly.

The emphasis throughout has been unashamedly 'organic', but interpretive theory has been included even where controversy exists: the theory of nuclear magnetic resonance is particularly satisfying and logical when treated semiempirically, but infrared theory is often conflicting in its predictions, mass spectroscopy theory is often speculative and electronic theory can be very mathematical. These strengths and weaknesses are emphasised throughout.

Students of chemistry, biochemistry or pharmacy at university or college will hopefully find the book easy to read and understand: the examples chosen for illustration are all simple organic compounds, and chapter 6 includes problems at an equally introductory level (so that students can *succeed* in problem-solving!) It is likely that the chapter supplements will be studied by students at honours chemistry degree or postgraduate level, and on the whole this is reflected in degree of complexity.

The author is reluctant to admit it, but he graduated at a time when no spectroscopic technique (other than X-ray crystallography) was taught in the undergraduate curriculum. He hopes that his own need to learn has given him a sympathetic insight into those dark areas that students find difficult to understand, and that the treatment accorded them in this book reflects their travail. His own colleagues have been of immense support, and did not laugh when he sat down to play. He thanks them all for it.

Heriot-Watt University   WILLIAM KEMP
Edinburgh
January 1975

1

ENERGY AND THE ELECTROMAGNETIC SPECTRUM

When the sun's rays are scattered in a rainbow or in a prism, the white light is separated into its constituent colours, or spectrum. The spectrum of visible light is a minute part of a much larger whole, called the *electromagnetic spectrum*: why 'electromagnetic'?

Visible light is a form of energy, which can be described by two complementary theories: the wave theory and the corpuscular theory. Neither of these theories alone can completely account for all the properties of light: some properties are best explained by the wave theory, and others by the corpuscular theory. The wave theory most concerns us here, and we shall see that the propagation of light by light waves involves both electric and magnetic forces, which gives rise to their common class name *electromagnetic radiation*.

1.1 UNITS

We can represent a light wave travelling through space by a sinusoidal trace as in figure 1.1. In this diagram λ is the *wavelength* of the light; different colours of light have different values for their wavelengths, so that, for example, red light has wavelength ≈ 800 nm, while violet light has wavelength ≈ 400 nm.

If we know the wavelength λ, we can calculate the inverse of this, $1/\lambda$,

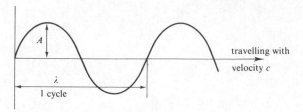

Figure 1.1 *Wave-like propagation of light (λ = wavelength, A = amplitude, c = 2.998 × 10^8 m s^{-1}).*

which is the number of waves per unit of length. This is most frequently used as the number of waves per cm, and is called the *wavenumber* $\bar{v}$ (cm^{-1}).

Also, provided we know the *velocity* with which light travels through space (c = 2.998 × 10^8 m s^{-1}) we can calculate the number of waves per second as the *frequency* of the light $v = c/\lambda$ (s^{-1}).

In summary, we can describe light of any given 'colour' by quoting either its wavelength λ, or its wavenumber $\bar{v}$, or its frequency v.

To take a numerical example, let us consider infrared light of wavelength $\lambda = 10^{-6}$ m $= 10^{-4}$ cm. For this light the wavenumber value is $\bar{v} = 1/10^{-4}$ cm $= 10^4$ cm^{-1}, and its frequency is $v = (3 \times 10^8$ m s$^{-1})/10^{-6}$ m $= 3 \times 10^{14}$ s^{-1}. Frequency units, in s^{-1}, are normally quoted in hertz (Hz) where 1 Hz = 1 cycle s^{-1}.

Unfortunately, users of the different spectroscopic techniques we shall meet in this book do not all use the same units, although it would be possible for them to do so. In some techniques the common unit is wavelength, in other techniques most workers use wavenumber, while in others we find frequency is the unit of choice. This is merely a question of custom and usage, but it makes comparison among the techniques a little less clear than it might be.

1.2 THE ELECTROMAGNETIC SPECTRUM

The sensitivity limits of the human eye extend from violet light (λ = 400 nm, 4×10^{-7} m) through the rainbow colours to red light (λ = 800 nm, 8×10^{-7} m). Wavelengths shorter than 400 nm and longer than 800 nm exist, but they cannot be detected by the human eye. *Ultraviolet* light (λ < 400 nm) can be detected on photographic film or in a photoelectric cell, and *infrared* light (λ > 800 nm) can be detected either photographically or using a heat detector such as a thermopile.

Beyond these limits lies a continuum of radiation, which is shown in figure 1.2. Although all of the different divisions have certain properties in common (all possess units of λ, v, $\bar{v}$, etc.) they are sufficiently different to require different handling techniques. Thus visible light (together with

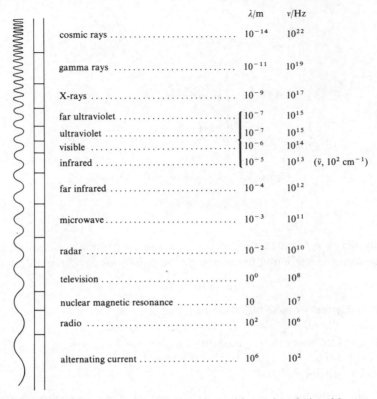

	λ/m	v/Hz
cosmic rays	10^{-14}	10^{22}
gamma rays	10^{-11}	10^{19}
X-rays	10^{-9}	10^{17}
far ultraviolet	10^{-7}	10^{15}
ultraviolet	10^{-7}	10^{15}
visible	10^{-6}	10^{14}
infrared	10^{-5}	10^{13} $(\bar{v}, 10^2\ cm^{-1})$
far infrared	10^{-4}	10^{12}
microwave	10^{-3}	10^{11}
radar	10^{-2}	10^{10}
television	10^{0}	10^{8}
nuclear magnetic resonance	10	10^{7}
radio	10^{2}	10^{6}
alternating current	10^{6}	10^{2}

Figure 1.2   *The electromagnetic spectrum, with wavelengths λ and frequency v shown.*

ultraviolet and infrared) can be transmitted in the form of 'beams', which can be bent by reflection or by diffraction in a prism; X-rays can pass through glass and muscle tissue and can be deflected by collision with nuclei; microwaves are similar to visible light, and are conducted through tubes or 'waveguides'; radiowaves can travel easily through air, but can also be conducted along a metal wire; alternating current travels only with difficulty through air, but easily along a metal conductor.

Alternating current is familiar as a wave-like electrical phenomenon, setting up fluctuating electrical fields as it travels through space or along a conductor. Associated with these *electric vectors* (and at 90° to them) are *magnetic vectors*. This relates to the simple experiment of placing a compass needle near a current-carrying conductor; the fluctuating electrical forces generate magnetic forces which deflect the compass needle. The relationship between these two quite different forms of energy is shown in figure 1.3. Alternating current is an electromagnetic phenomenon; all other parts of the spectrum in figure 1.2 possess electric and magnetic vectors and the name *electromagnetic radiation* is given to all the energy forms of this genre.

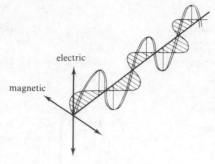

electric

magnetic

Figure 1.3 *Propagation of alternating electric forces and the related magnetic fields.*

The energy associated with regions of the electromagnetic spectrum is related to wavelength and frequency by the equations

$$E = hv = hc/\lambda$$

where E = energy of the radiation in joules/J,
    h = Planck's constant/6.626×10^{-34} J s,
    v = frequency of the radiation/Hz,
    c = velocity of light/2.998×10^8 m s^{-1},
    λ = wavelength/m.

The higher the frequency, the higher the energy; the longer the wavelength, the lower the energy. Cosmic radiation is of very high energy; ultraviolet light is of higher energy than infrared light, etc.

Most references to energy will be expressed in joules (J), but the electron volt will be mentioned in mass spectroscopy; 1 electron volt, 1 eV $\approx 1.6021 \times 10^{-9}$ J, so that ultraviolet light of wavelength 100 nm has an energy of about 12 eV.

To express energy in terms of J mol^{-1}, the expressions $E = hv$, etc., must be multiplied by the Avogadro constant $N_A (= 6.02 \times 10^{23}$ mol^{-1}). For example, ultraviolet light of wavelength 200 nm has an energy $E = hv = hc/\lambda \approx 10^{-18}$ J. When multiplied by N_A, this can be expressed as $\approx 6 \times 10^5$ J mol^{-1}, or 600 kJ mol^{-1}

1.3   ABSORPTION OF ELECTROMAGNETIC RADIATION BY ORGANIC MOLECULES

If we pass light from an ultraviolet lamp through a sample of an organic molecule such as benzene, some of the light is absorbed. In particular, some of the wavelengths (frequencies) are absorbed and others are virtually unaffected.

We can plot the changes in absorption against wavelength as in figure 1.4 and produce an *absorption spectrum*. The spectrum presented in figure 1.4 shows *absorption bands* at several wavelengths, for example 255 nm.

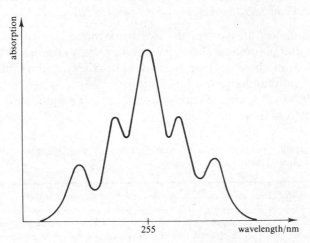

Figure 1.4   *An absorption spectrum.*

The organic molecule is absorbing light of $\lambda = 255$ nm, which corresponds to energy absorption of 470 kJ mol^{-1}. Energy of this magnitude is associated with changes in the electronic structure of the molecule, and when a molecule absorbs this wavelength, electrons are promoted to higher energy orbitals as represented in figure 1.5. The energy transition $E_1 \rightarrow E_2$ corresponds to the absorption of energy *exactly* equivalent to the energy of the wavelength absorbed

$$\Delta E = (E_2 - E_1) = hc/\lambda = h\nu$$

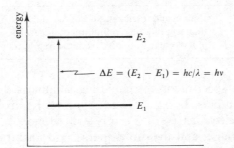

Figure 1.5   *Energy transition for the absorption of light or other electromagnetic radiation.*

While this example refers specifically to ultraviolet light, the same principle holds for the absorption of energy from any part of the electromagnetic spectrum.

A molecule can only absorb a particular frequency, if there exists within the molecule an energy transition of magnitude $\Delta E = h\nu$.

Although almost all parts of the electromagnetic spectrum are used for studying matter, in organic chemistry we are mainly concerned with energy absorption from three or four regions—ultraviolet and visible, infrared, microwave and radiofrequency absorption.

Table 1.1 shows the kind of information that can be deduced from studying the absorption of these radiations.

Table 1.1   Summary of spectroscopic techniques in organic chemistry and the information obtainable from each

Radiation absorbed	Effect on the molecule (and information deduced)
ultraviolet–visible λ, 190–400 nm and 400–800 nm	changes in electronic energy levels within the molecule (extent of π-electron systems, presence of conjugated unsaturation, and conjugation with nonbonding electrons)
infrared λ, 2.5–25 μm $\bar{\nu}$, 400–4000 cm^{-1}	changes in the vibrational and rotational movements of the molecule (detection of functional groups, which have specific vibration frequencies, for example C=O, NH$_2$, OH, etc.)
microwave ν, 9.5×10^9 Hz	electron spin resonance or electron paramagnetic resonance; induces changes in the magnetic properties of unpaired electrons (detection of free radicals and the interaction of the electron with, for example, nearby protons)
radiofrequency ν, 60–300 MHz	nuclear magnetic resonance; induces changes in the magnetic properties of certain atomic nuclei, notably that of hydrogen (hydrogen atoms in different environments can be detected and counted, etc.)
electron-beam impact 70 eV, 6000 kJ mol^{-1}	ionisation and fragmentation of the molecule into a spectrum of fragment ions (determination of molecular weight and deduction of molecular structures from the fragments produced)

The last of the spectroscopic techniques summarised in table 1.1 is different from the others. In *mass spectroscopy* we bombard the molecule with high-energy electrons ($\approx$ 70 eV, or 6000 kJ mol^{-1}), and cause the molecule first to ionise, and then to disperse into an array (or spectrum) of fragment ions of different masses. This *mass spectrum* presents us with a jigsaw pattern of fragments from which we have to reconstruct a picture of the whole molecule.

FURTHER READING

C. J. Cresswell, O. Runquist and M. M. Campbell, *Spectral Analysis of Organic Compounds*, Burgess, Minneapolis (2nd edn 1972). (Excellent programmed text for self-teaching).

R. J. Taylor, *The Physics of Chemical Structure*, Unilever Educational Booklets: Advanced Series, Unilever, London (1969). (Available free.)

C. N. Banwell, *Fundamentals of Molecular Spectroscopy*, McGraw-Hill, New York (1966).

G. M. Barrow, *Introduction to Molecular Spectroscopy*, McGraw-Hill, New York (1962). (These last two books dwell more on physical interpretations and introduce the mathematics of spectroscopy.)

2

INFRARED SPECTROSCOPY†

When infrared light is passed through a sample of an organic compound, some of the frequencies are absorbed while other frequencies are transmitted through the sample without being absorbed. If we plot the per cent absorbance or per cent transmittance against frequency, the result is an infrared spectrum.

Figure 2.1 is the infrared spectrum of a mixture of long-chain alkanes (liquid paraffin, or Nujol), showing that *absorption bands* appear in the regions around 3000 cm^{-1} and 1400 cm^{-1}; other frequencies do not interact with the sample and are consequently almost wholly transmitted.

The alkane molecules will only absorb infrared light of a particular frequency if there is an energy transition within the molecule such that $\Delta E = h\nu$. The transitions involved in infrared absorption are associated with *vibrational* changes within the molecule; for example, the band near 3000 cm^{-1} (that is, corresponding to 9.3×10^{13} Hz) has exactly the same frequency as a C—H bond undergoing *stretching vibrations*.

The absorption band near 3000 cm^{-1} is therefore called the C—H (*stretch*) *absorption*, usually represented as C—H *str*.

The bands around 1400 cm^{-1} correspond to the frequency of the *bending vibrations* of C—H bonds, and are called the C—H (*bend*) *absorptions*.

† The infrared spectra reproduced in this chapter were recorded on a Perkin–Elmer Model 700 infrared spectrophotometer, with the exception of figure 2.10, which was recorded on a Perkin–Elmer Model 137 spectrometer. Figure 2.23 is reproduced with permission of Perkin–Elmer Ltd, Beaconsfield, Buckinghamshire. The correlation charts on pages 40–51 are reproduced with permission from *Qualitative Organic Analysis* by Kemp, McGraw-Hill, Maidenhead (1970).

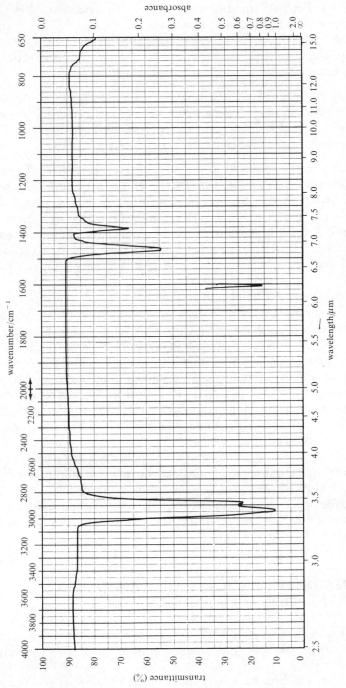

Figure 2.1 *Infrared spectrum of mixed long-chain alkanes (liquid paraffin, Nujol). Liquid film.*

Alternatively, bending vibrations are referred to as *deformations*, so that C—H deformation bands can be labelled as C—H *def*.

Infrared spectroscopy is therefore basically vibrational spectroscopy, and the principal value of the technique to organic chemists relates to the following observation.

> *Different bonds* (C—C, C=C, C≡C, C—O, C=O, O—H, N—H, *etc.*) *have different vibrational frequencies, and we can detect the presence of these bonds in an organic molecule by identifying this characteristic frequency as an absorption band in the infrared spectrum.*

For example, the spectrum in figure 2.3 (page 12) is that of a carbonyl compound, and we know that the strong band at 1700 cm^{-1} is associated with the stretching vibration of the C=O bond (thus we say C=O *str* appears at 1700 cm^{-1}, or $\nu_{C=O} = 1700$ cm^{-1}). Similarly, by examining the spectrum in figure 2.17 (page 65) we can say that the compound contains a nitrile group; the strong band at 2250 cm^{-1} is the C≡N stretch absorption. (Alternatively we say C≡N *str* appears at 2250 cm^{-1}, or $\nu_{C≡N} = 2250$ cm^{-1}.)

By examining a large number of compounds of a given class, we can draw up data in the form of tables or charts, which allow us to correlate the presence of absorption at a particular frequency with the presence of a functional group within the molecule. Such a set of *correlation charts* appears on pages 40–51, and two simple examples will illustrate their use.

The infrared spectrum in figure 2.2 is known to be that of an alkyne: the spectrum is examined in conjunction with chart 1, and we can see that it shows strong absorptions corresponding to C≡C *str* (at 2150 cm^{-1}) *and* to ≡C—H *str* (at 3320 cm^{-1}); this latter proves that the alkyne has a terminal triple bond. (The example is 1-octyne.)

Using chart 2, we can distinguish acetophenone from benzaldehyde, because although the spectra of benzaldehyde (figure 2.3) and acetophenone (figure 2.4) both show C=O *str* absorptions (around 1700 cm^{-1}), the benzaldehyde spectrum also shows the characteristic aldehyde C—H *str* absorptions (near 2800 cm^{-1}), which are absent from the acetophenone spectrum.

2.1   UNITS OF FREQUENCY, WAVELENGTH AND WAVENUMBER

The position of an absorption band can be specified in units of frequency, ν (s^{-1}, or Hz); or wavelength, λ (micrometres, μm), or wavenumber, $\bar{\nu}$ (reciprocal centimetres, cm^{-1}).

The stretching absorption of C—H bonds therefore appears at $\approx 9.3 \times 10^{13}$ s^{-1} ≡ 9.3×10^{13} Hz ≡ 3.3 μm ≡ 3000 cm^{-1}.

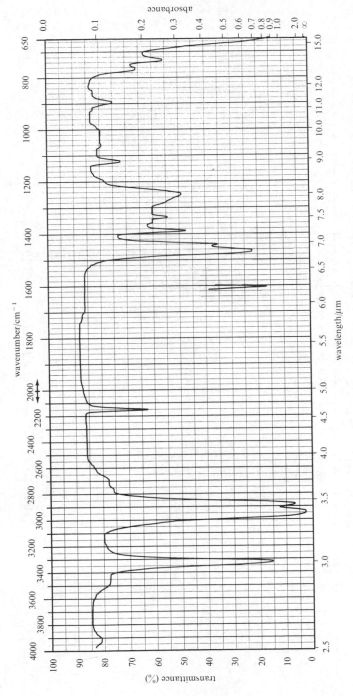

Figure 2.2 *Infrared spectrum of 1-octyne (CH₃(CH₂)₅C≡CH). Liquid film.*

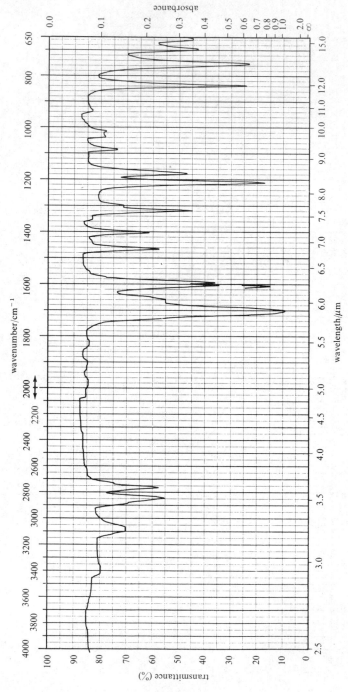

Figure 2.3 *Infrared spectrum of benzaldehyde* (PhCHO). *Liquid film.*

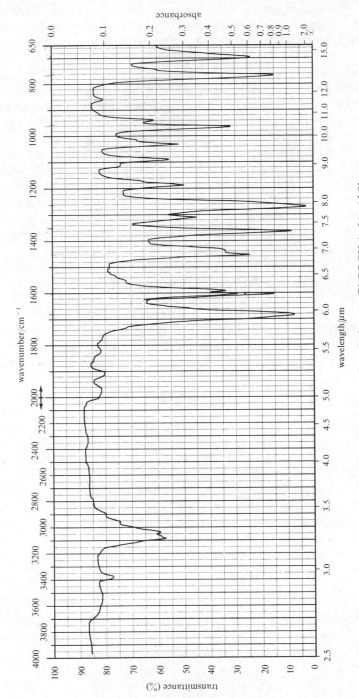

Figure 2.4 *Infrared spectrum of acetophenone (PhCOCH₃). Liquid film.*

The majority of chemists use wavenumber units (cm^{-1}), although a very small minority uses wavelength (μm). True frequencies are virtually never used (which is unfortunate, since only by quoting frequency can one obtain a picture of the C—H bond stretching and contracting 9.3×10^{13} times per second).

We shall occasionally commit the universal sin of referring to a vibration as having a 'frequency' of x cm^{-1} (rather than a wavenumber of x cm^{-1}). To avoid the confusion between v and $\bar{v}$ is commendable, but the syntax is usually clumsy.

In the discussions that follow, wavenumber units are favoured, although spectra and chart data show both wavenumber and wavelength units.

Organic applications of infrared spectroscopy are almost entirely concerned with the range 650–4000 cm^{-1} (15.4–2.5 μm). The region of frequencies lower than 650 cm^{-1} is called the *far infrared*, and that of frequencies higher than 4000 cm^{-1} is called the *near infrared*. These regions are respectively farther from and nearer to the visible spectrum.

The far infrared contains a few absorptions of interest to organic chemists, notably carbon–halogen bond absorptions, and the absorptions associated with rotational changes within molecules.

The near infrared reaches right to the long wavelength limit of the visible spectrum (0.75 μm), and mainly shows absorptions that are harmonic overtones of the fundamental vibrations found within the 'normal' range. Little use has been made of these absorptions by organic chemists.

2.2 MOLECULAR VIBRATIONS

At ordinary temperatures, organic molecules are in a constant state of vibration, each bond having its characteristic stretching and bending frequency, and being capable of absorbing light of that frequency. The vibrations of two atoms joined together by a chemical bond can be likened to the vibrations of two balls joined by a spring: using this analogy, we can rationalise several features of infrared spectra. For example, to stretch a spring requires more energy than to bend it; thus the stretching energy of a bond is greater than the bending energy, and *stretching absorptions of a bond appear at higher frequencies in the infrared spectrum than the bending absorptions of the same bond.*

2.2.1 CALCULATION OF VIBRATIONAL FREQUENCIES
We can calculate the vibrational frequency of a bond with reasonable accuracy, in the same way as we can calculate the vibrational frequency of a ball and spring system; the equation is Hooke's law

$$v = \frac{1}{2\pi}\left(\frac{k}{(m_1 m_2 / m_1 + m_2)}\right)^{1/2}$$

where v = frequency,

   k = a constant related to the strength of the spring (the *force constant* of the bond),

m_1, m_2 = the masses of the two balls (or atoms).

The quantity $m_1 m_2 / (m_1 + m_2)$ is often expressed as μ, the *reduced mass* of the system.

As an example, we can calculate the approximate frequency of the C—H stretching vibration from the following data

$$k = 500 \text{ N m}^{-1} = 5.0 \times 10^5 \text{ g s}^{-2} \text{ (since 1 newton} = 10^3 \text{ g m s}^{-2})$$

$$m_C = \text{mass of the carbon atom} = 20 \times 10^{-24} \text{ g}$$

$$m_H = \text{mass of the hydrogen atom} = 1.6 \times 10^{-24} \text{ g}$$

$$v = \frac{7}{2 \times 22} \left(\frac{5.0 \times 10^5 \text{ g s}^{-2}}{(20 \times 10^{-24} \text{ g})(1.6 \times 10^{-24} \text{ g})/(20 + 1.6)10^{-24} \text{ g}} \right)^{1/2}$$

$$= 9.3 \times 10^{13} \text{ s}^{-1}$$

To express this in wavenumbers ($\bar{v}$) we use the relationship shown in section 1.2 (where c is the velocity of light $= 3.0 \times 10^8 \text{ m s}^{-1}$)

$$\bar{v} = \frac{v}{c} = \frac{9.3 \times 10^{13} \text{ s}^{-1}}{3.0 \times 10^8 \text{ m s}^{-1}}$$

$$= 3.1 \times 10^5 \text{ m}^{-1}$$

$$= 3100 \text{ cm}^{-1}$$

It is important qualitatively to restate the principles embodied in these calculations.

The vibrational frequency of a bond is expected to increase when the bond strength increases, and also when the reduced mass of the system decreases.

We can then predict that C=C and C=O *str* will have higher frequencies than C—C and C—O *str*, respectively: we also expect to find C—H and O—H *str* absorptions at higher frequencies than C—C and C—O *str*. Similarly we would predict O—H *str* to be of higher frequency than O—D *str*.

Without accurate data on force constants, however, some caution should be exercised in predicting exact trends other than in this general way.

For example, on the basis of mass, we expect X—H *str* frequencies to fall along the series C—H, N—H, O—H, F—H; in fact they rise, mainly due to increasing electronegativity. There are also extreme circumstances in which O—H *str* has lower frequency than O—D *str*. Section 2.3 deals with other factors influencing vibrational frequencies.

2.2.2   MODES OF VIBRATION

Molecules with large assemblages of atoms possess very many vibrational frequencies; for a nonlinear molecule with n atoms, the number of vibrational modes is $(3n - 6)$, so that methane theoretically possesses 9, and ethane has 18. Does this lead to methane having 9 (and not more than 9) absorption bands in the infrared?

Figure 2.5 shows that for a single methylene group several vibrational modes are available, and any atom joined to two other atoms will undergo comparable vibrations (for example any AX_2 system such as NH_2, NO_2).

STRETCHING MODES FOR CH_2

symmetric   antisymmetric

BENDING OR DEFORMATION MODES

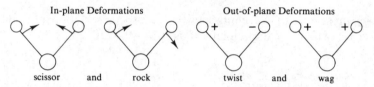

In-plane Deformations   Out-of-plane Deformations

scissor   and   rock   twist   and   wag

Figure 2.5   *Vibration modes in methylene groups. Other AX_2 groups behave similarly, and methyl groups behave analogously.*

Each of the different vibration modes may (and frequently does) give rise to a different absorption band, so that CH_2 groups give rise to two C—H *str* bands at v_{symm} and v_{anti}. Other vibrations may not give rise to absorption, since some may have frequencies outside the normal infrared region being examined. Some of the vibrations may have the same frequency (that is, they are degenerate) and their absorption bands will overlap. It is also necessary, in order to 'see' an absorption band, that the particular vibration should produce a fluctuating dipole (and thus a fluctuating electric field) otherwise it cannot interact with the fluctuating electric fields of the infrared light. Thus the stretching of a symmetrically substituted bond (for example C≡C in acetylene) produces no change in the dipole of the system, and therefore this vibration cannot interact with infrared light. (See Raman spectroscopy, section 2S.3).

In addition to these *fundamental vibrations* thus far discussed, other frequencies can be generated by modulation, etc. of the fundamentals. *Overtone bands* (harmonics) appear at integer multiples of fundamental vibrations, so that strong absorptions at, say, $800\ cm^{-1}$ and $1750\ cm^{-1}$

will also give rise to weaker absorptions at $1600 \, \text{cm}^{-1}$ and $3500 \, \text{cm}^{-1}$ respectively. Two frequencies may interact to give *beats*, which are *combination* or *difference* frequencies; thus absorptions at $x \, \text{cm}^{-1}$ and $y \, \text{cm}^{-1}$ interact to produce two weaker beat frequencies at $(x \pm y) \, \text{cm}^{-1}$.

Modulation by addition or subtraction of two frequencies has important consequences in Raman spectroscopy (see section 2S.3).

2.2.3 QUANTUM RESTRICTIONS
When a vibrating ball and spring system gains energy, the *frequency* of vibration does not alter, but the *amplitude* increases. Thus, in the stretching of a spring, increased energy leads to greater degrees of extension and contraction of the spring. While this is also true of the vibrations of inter-atomic bonds, quantum theory applies to chemical bonds and imposes certain additional restrictions. Vibrational energy can only increase by quantum jumps, so that the energy difference between successive vibrational levels is the familiar $\Delta E = h\nu$ or $\Delta E = N_A h\nu$ (see section 1.3).

Absorption of infrared light around $3000 \, \text{cm}^{-1}$ (corresponding to C—H *str*) involves an increase in the energy of the molecule of $N_A \, h\nu$, or about $37 \, \text{kJ mol}^{-1}$. Once in the vibrationally excited state, the molecule can give up this extra energy by rotational, collision or translational (kinetic) processes, etc.

2.3 FACTORS INFLUENCING VIBRATIONAL FREQUENCIES

Many factors influence the precise frequency of a molecular vibration, and it is usually impossible to isolate one effect from another. For example, the C=O *str* frequency in the ketone $RCOCH_3$ is lower than in $RCOCl$; is the change in frequency of the C=O *str* due to the difference in *mass* between CH_3 and Cl, or is it associated with the *inductive* or *mesomeric* influence of Cl on the C=O bond; perhaps there is some *coupling* interaction between the C=O and C—Cl bonds, or is there some steric effect which alters the *bond angles*?

We shall discuss here frequency shifts, which are brought about by structural changes in the molecule, or by interaction between functional groups. Due emphasis will be placed on those features that are most valuable in explaining the characteristic appearance and positions of the group frequencies.

Primary mass effects (for example, the mass effect of changing C—H to C—Cl) have been mentioned in section 2.2; secondary mass effects (for example, the effect on C=O *str* of changing CO—CH_3 to CO—Cl) are very difficult to study because of the unavoidable intrusion of electronic effects. Frequency shifts also take place on moving from condensed phases to dilute solutions, as mentioned in the section on sampling techniques (see section 2.5).

2.3.1 VIBRATIONAL COUPLING

An isolated C—H bond has only one stretching frequency, but the stretching vibrations of C—H bonds in CH_2 groups combine together to produce two *coupled vibrations* of different frequencies—the antisymmetric, v_{anti}, and symmetric, v_{symm}, combinations discussed earlier and illustrated in figure 2.5. The equivalent coupled vibrations of CH_3 groups are of different frequencies from those of CH_2 groups, and all four vibrations can be seen in high-resolution spectra of compounds containing both CH_2 and CH_3 groups.

Vibrational coupling takes place between two bonds vibrating with similar frequency provided the bonds are reasonably close in the molecule; the coupling vibrations may *both be fundamentals* (as in the coupled stretching vibrations of AX_2 groups) or a *fundamental vibration may couple with the overtone* of some other vibration. This latter coupling is frequently called Fermi resonance, after Enrico Fermi who first described it.

Vibrational coupling is a feature of other AX_2 groups, so that the functions listed in table 2.1 exhibit, not one, but two stretching bands—antisymmetric and symmetric A—X *str* (antisymmetric usually being of higher frequency).

Carboxylic acid anhydrides. These give rise to two C=O *str* absorptions, v_{anti} and v_{symm} (around 1800–1900 cm^{-1}, with a separation of about 65 cm^{-1}); coupling occurs between the two carbonyl groups, which are *indirectly* linked through —O—: the interaction is presumably encouraged because of the slight double-bond character in the carbonyl–oxygen bonds brought about by resonance, since this will keep the system coplanar. The high-frequency band in this case is the symmetric C=O *str*.

Table 2.1 Typical antisymmetric and symmetric stretching frequencies for common AX_2 groups

Group	Antisymmetric v_{anti}/cm^{-1}	Symmetric v_{symm}/cm^{-1}
—CH$_2$—	3000	2900
—NH$_2$	3400	3300
—NO$_2$ $\left(-N\begin{smallmatrix}O\\O\end{smallmatrix}\right)$	1550	1400
—SO$_2$— $\left(-\overset{O}{\underset{O}{S}}-\ \text{or}\ -\overset{O}{\underset{O}{S}}-O-\right)$	1350	1150
$-C\begin{smallmatrix}O\\O\end{smallmatrix}^-$ (salts of $-C\begin{smallmatrix}O\\OH\end{smallmatrix}$)	1600	1400

anhydrides—resonance forms

Amides. These show two absorption bands around 1600–1700 cm^{-1} corresponding mainly to C=O *str* and N—H *def*, but because of vibrational coupling the original character of the vibrations are modified. The two bands are not pure C=O *str* and N—H *def*, and are usually referred to as the amide I and amide II bands. Amide I may be as high as 80 per cent C=O *str* in character, but amide II is a strongly coupled interaction between N—H *def* and C—N *str*. (See also section 2.9.)

In *aldehydes* the C—H *str* absorption usually appears as a doublet because of interaction between the C—H *str* fundamental and the overtone of C—H *def*.

2.3.2 HYDROGEN BONDING
Hydrogen bonding, especially in O—H and N—H compounds, gives rise to a number of effects in infrared spectra, and its importance here can scarcely be overemphasised. While most routine organic work will involve relatively nonassociating solvents (CCl$_4$, CS$_2$, CHCl$_3$), more polar solvents such as acetone or benzene will certainly influence O—H and N—H absorptions. Carbonyl groups or aromatic rings *in the same molecule* as the O—H or N—H group may cause similar shifts by intramolecular action.

Alcohols and phenols. Figure 2.6 shows the infrared spectrum of an alcohol (1-butanol) recorded as a liquid film; the dotted line insert around 3500 cm^{-1} was recorded in dilute solution (about 1 per cent in CCl$_4$). At low concentrations a sharp band appears at 3650 cm^{-1} in addition to the broad band at 3350 cm^{-1}.

The sharp band is O—H *str* in *free* alcohol molecules, the broad band is O—H *str* in hydrogen-bonded alcohol molecules.

Alcohols and phenols in condensed phases (bulk liquid or KBr discs, etc.) are strongly hydrogen bonded, usually in the form of a dynamic polymeric association; dimers, trimers and tetramers also exist, and this leads to a wide envelope of absorptions and hence to broadening of the absorption band. In dilute solution in inert solvents (or in the vapour phase) the proportion of free molecules increases and these give rise to the 3650 cm^{-1} band.

Is it reasonable that bonded O—H *str* should appear at lower frequency than free O—H *str*?

The hydrogen bond can be regarded as a resonance hybrid of I and II (approximating overall to III) so that hydrogen bonding involves a lengthening of the original O—H bond. This bond is consequently weakened (that is, its force constant is reduced), so the stretching frequency is lowered.

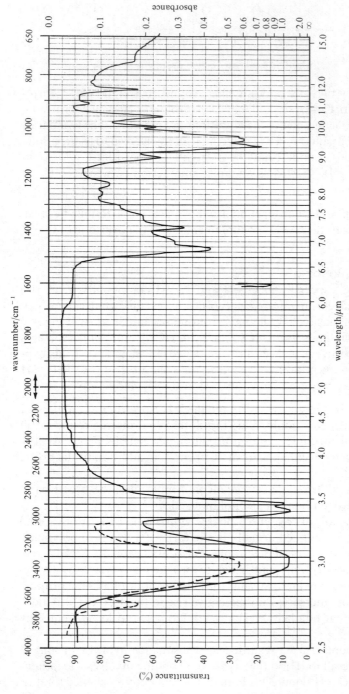

Figure 2.6 *Infrared spectrum of 1-butanol* ($CH_3CH_2CH_2CH_2OH$). *Complete spectrum—liquid film: dotted line insert near 3500 cm^{-1}—dilute solution in carbon tetrachloride.*

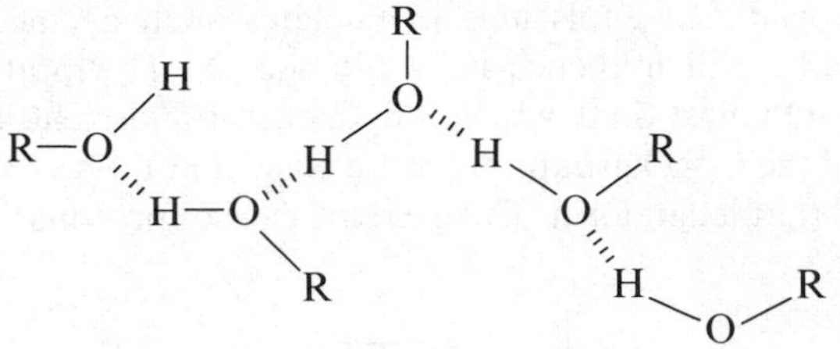

polymeric association of O—H compounds

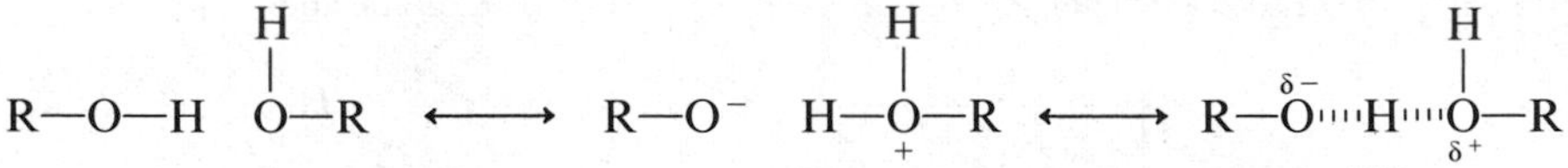

$$R-O-H \quad O-R \overset{H}{\underset{}{\longleftrightarrow}} R-O^- \quad H-\overset{H}{\underset{+}{O}}-R \longleftrightarrow R-\overset{\delta-}{O}\cdots H\cdots\overset{\delta+}{\underset{}{O}}-R$$

$$\text{I} \qquad\qquad\qquad \text{II} \qquad\qquad\qquad \text{III}$$

lengthening of O—H bond in hydrogen bonding

Enols and chelates. Hydrogen bonding in enols and chelates is particularly strong, and the observed O—H *str* frequencies may be very low (down to 2800 cm^{-1}). Since these bonds are not easily broken on dilution by an inert solvent, free O—H *str* may not be seen at low concentrations.

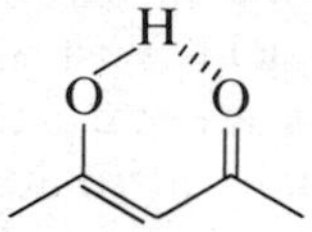

enol
$(CH_3COCH_2COCH_3)$

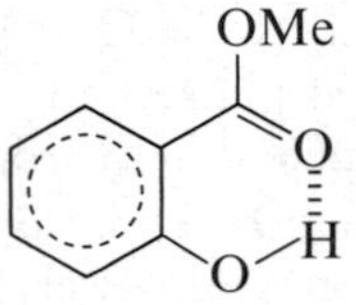

chelate
(methyl salicylate)

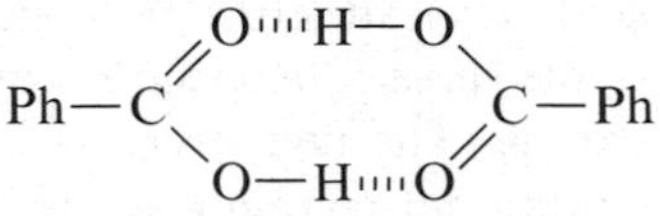

dimer
(of benzoic acid)

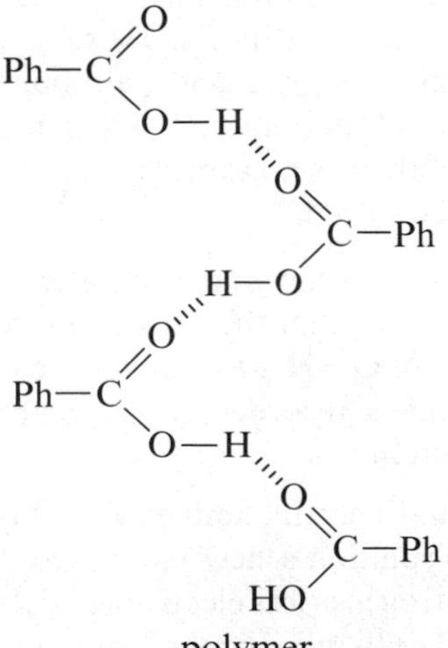

polymer
(of benzoic acid)

Carbonyl compounds. In enols and in chelates such as methyl salicylate hydrogen bonding will influence not only the O—H vibration frequency but also the C=O vibration to which it hydrogen-bonds. The key factor here is the basicity of the C=O group: the more basic it is, the stronger will be the hydrogen bond that it can form. The extreme case of protonation shows that

hydrogen bonding

protonation

the C=O bond has increased single-bond character and longer length: the same tendency occurs in hydrogen bonding, leading to a lowering of the vibration frequency.

Carboxylic acids. Figure 2.7 shows the infrared spectrum of benzoic acid, and the exceedingly broad band reaching from 2500–3500 cm^{-1} is hydrogen-bonded O—H *str*. We are seeing here the O—H *str* band for the carboxylic acid *dimer* structure: in condensed phases, all carboxylic acids exist in this stable *dimeric association* in which the hydrogen bonds are particularly strong. (The fine structure on the O—H *str* peak is usually attributed to vibrational coupling with overtones of lower frequencies.) In very dilute solution in hexane it is just possible to distinguish free O—H *str*, but this is extreme dilution. Even in CCl$_4$ some degree of hydrogen bonding to solvent arises and in extreme dilution in CCl$_4$ the free O—H *str* absorption is seen at lower frequency than in hexane. Polymeric association is also known to occur in carboxylic acids, although dimeric association is the norm; the proportion of monomer to dimer increases in solvents such as benzene, and in dioxan there is no dimer formed, since the acid hydrogen-bonds preferentially to the solvent.

π-Cloud interactions. Since alkene and aromatic π-bonds can behave as Lewis bases it is not surprising that they can form hydrogen bonds to acidic hydrogens; the frequency of O—H *str* in phenols can be lowered by 40–100 cm^{-1} when the spectrum is recorded in benzene solution compared to carbon tetrachloride solution.

Amines. In condensed-phase spectra, amines show bonded N—H *str* around 3300 cm^{-1} and in dilute solution a new band near 3600 cm^{-1} corresponds to free N—H *str*. Since nitrogen is less electronegative than oxygen, hydrogen bonds in amines are weaker than in alcohols, and the shifts in frequency are also correspondingly less dramatic than in alcohols.

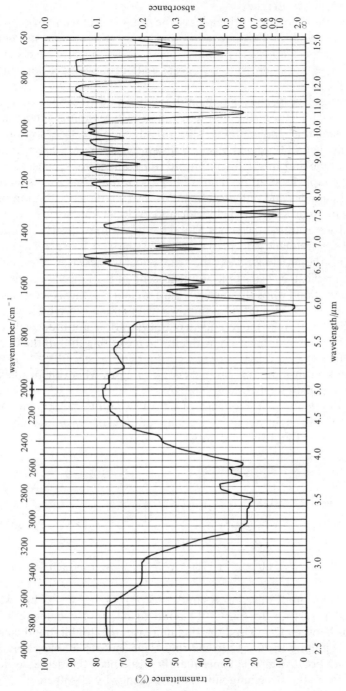

Figure 2.7 Infrared spectrum of benzoic acid (PhCO₂H). KBr disc.

2.3.3   ELECTRONIC EFFECTS

One can use the theoretical principles of the organic chemist to explain many of the frequency shifts that occur in vibrations when the substituents are altered. The expected *inductive* and *mesomeric* (or *resonance*) effects are seen to be at work, together with an occasional through-space influence (or *field effect*).

Many unresolved problems remain, however, and we cannot concentrate on successes and ignore the many instances where simple theory fails to offer a reasonable explanation. Vibrational coupling (see section 2.3.1) often means that an observed absorption band is *not* purely associated with one bond alone and this will complicate our explanations; most C—H *def* modes are coupled vibrations, and we have seen that C=O *str* is a coupled vibration in amides and anhydrides, and is also a coupled vibration in such simple compounds as benzoyl chloride and cyclopentanone.

Again, if we examine the series MeOH, PhOH, MeCOOH, we find that the O—H *str* frequency decreases, while in the series MeNH$_2$, PhNH$_2$, MeCONH$_2$, the N—H *str* frequency inexplicably increases. We must also consider the effect of electronic influences on the strengths of the bonds *adjacent* to the bond whose frequency we are measuring: thus, pictorially, if we stiffen up the bonds to a C=O group, we will make it more difficult for the carbonyl carbon atom to move, and all of the vibration amplitude will have to be taken up by the oxygen atom, with almost inevitable shift in the C=O *str* frequency (see section 2.3.4).

With caution in mind, we can now look at cases where theory has been successful in explaining frequency shifts.

Conjugation lowers the frequency of C=O *str* and C=C *str*, whether the conjugation is brought about by αβ-unsaturation or by an aromatic ring. Compare I with II and III; or compare IV with V.

I   II   III   IV   V

$v_{C=O} \approx 1720$ cm^{-1} 1700 cm^{-1} 1700 cm^{-1} $v_{C=C} \approx 1650$ cm^{-1} 1610 cm^{-1}

The explanation of this shift is similar for C=O and C=C but we shall illustrate it in relation to the C=O bond in III. In III, delocalisation of π-electrons between C=O and the ring increases the double-bond character of the bond joining the C=O to the ring. This leads to a lower bond order in the C=O bond, which is consequently weakened; the decrease in force constant lowers the stretching vibration frequency by 20–30 cm^{-1}.

One can also attribute such C=O frequency shifts to the mesomeric

(or resonance) effect: any substituent that enhances the mesomeric shift
will decrease the bond order of the C=O bond and lead to lower C=O *str*

frequency. Conjugation with phenyl (in VII) does so, and a $+M$ group
such as *p*-MeO in VIII will lead to even lower frequencies. A *p*-NO$_2$ group
$(-M)$ will oppose these trends and lead to higher frequencies (as in IX).

Inductive effects are difficult to consider in isolation from mesomeric
effects: in some molecules I is more important than M, while in others the
reverse is true.

In amides, XI, the $+M$ effect produces a lengthening (weakening) of the
C=O bond, leading to lower frequency than in the corresponding ketone X:
the $-I$ effect of nitrogen is here being dominated by $+M$. In contrast, the $-I$
effect of chlorine in acyl chlorides, XII, is more influential than $+M$, and here
an opposite shift (to higher frequency) occurs.

Esters represent another example of the conflict between I and M effects.
In alkyl esters, XIII, the nonbonding electrons on oxygen increase the $+M$
conjugation, tending to lower the C=O frequency. The electronegativity of
oxygen, $-I$, operates in the opposite sense, but $+M$ is apparently dominant.
In phenyl esters XIV, however, the nonbonding electrons are partly drawn
into the ring, and their conjugation with C=O is consequently diminished.
When this happens, the $-I$ effect of oxygen becomes dominant, and C=O
moves to higher frequency.

In examples such as these it is easier to rationalise the shifts than it is to
predict them, and caution should be exercised in applying the rules to new
situations. The importance of vibrational coupling requires constant
restatement.

2.3.4   BOND ANGLES

In ketones, the correlation charts show that highest C=O frequencies arise
in the strained cyclobutanones, and we can explain this in terms of bond-
angular strain: the C—CO—C bond angle is reduced below the normal 120°
leading to increased s character in the C=O bond. The C=O bond is

shortened and therefore strengthened and so $v_{C=O}$ increases. If the bond angle is pushed outwards above 120°, the opposite effect operates, and for this reason di-*tert*-butyl ketone has a very low $v_{C=O}$ (1697 cm^{-1}).

An alternative view involves no change in the C=O force constant, but merely an increased rigidity in the C—CO—C bond system as ring size decreases: C=O stretching must in these circumstances couple more effectively with C—C stretching, leading to higher C=O *str* frequencies.

Cycloalkenes also show such an effect, but a less simple relationship holds. Thus in cycloalkenes XV $v_{C=C}$ falls with increasing strain, but reaches a minimum in cyclobutene. In cyclobutene XVI stretching of C=C involves only *bending* of the attached C—C bonds: in all the others (where the internal angles are not 90°) C=C stretching must involve some stretching of the adjacent C—C bonds, which involves increasing the energy (frequency) of C=C *str*.

$$\left(\overset{=}{\underset{(CH_2)_n}{}} \right)$$

$n = 1$ to 6	except	cyclobutene
XV		XVI
$v_{C=C} \approx$ 1610–1650 cm^{-1}		1566 cm^{-1}

C—H *stretching* vibrations move to higher frequency in the sequence alkane–alkene–alkyne. As hybridisation goes from sp^3–sp^2–sp, the s character of the C—H bond increases; bond lengths become shorter, and frequencies rise. Cyclopropanes have high C—H *str* frequencies for the same reason (typical values being 3040–3070 cm^{-1}): the C—$\widehat{C}$—C bond angle is substantially contracted below the normal 109.5°, leading to increased s character in the C—H bonds, and thus to higher frequencies.

2.3.5 FIELD EFFECTS

Two groups often influence each other's vibrational frequencies by a through-space interaction, which may be electrostatic and/or steric in nature. The best examples of this *field effect* are interactions between carbonyl groups and halogen atoms; for example in the α-chloroketone derivatives of steroids XVII, C=O *str* frequency is higher when Cl is equatorial than when it is axial. Presumably the nonbonding electrons of oxygen and chlorine undergo repulsion when they are close together in the molecule; this results in a change in the hybridisation state of oxygen, and therefore a shift in C=O *str* frequency.

In *o*-chlorobenzoic acid esters this field effect shifts the C=O frequency in the rotational isomer XVIII, and not in the isomer XIX; both isomers are normally present, so that *two* C=O *str* absorptions are observed in the spectrum of this compound.

XVII   XVIII   XIX

2.4 INSTRUMENTATION—THE INFRARED SPECTROMETER

The essential features of an infrared spectrometer are a source of infrared light, a monochromator and a detector. Light from the source is passed through an organic sample, split into its individual frequencies in the monochromator, and the relative intensities of the individual frequencies measured in the detector.

2.4.1 INFRARED SOURCES
Common sources are electrically heated rods of the following

'Nernst glower' (sintered mixtures of the oxides of Zr, Y, Er, etc.)
'Globar' (silicon carbide)
Various ceramic materials

Since the infrared output from these sources varies in intensity over the required frequency range, a compensating variable *slit* is programmed to open and close in unison with the scanning over the individual frequencies.

2.4.2 MONOCHROMATORS
Both prisms and gratings are used (prisms being largely obsolescent), the most common prism material being NaCl; since NaCl is only transparent down to 625 cm^{-1}, other metal halides must be used for low-frequency work (for example CsI, or a mixture of ThBr and ThI known as KRS-5). In general, gratings give better resolution at high frequency than do prisms, and NaCl suffers from the additional disadvantage of being hygroscopic so that the optics must be protected from condensation of moisture, usually by maintaining them at about 20°C above ambient temperature.

2.4.3 DETECTORS
Most modern instruments use thermopile detectors: these work on the thermocouple principle that if two dissimilar metal wires are joined head to tail, then a difference in temperature between head and tail causes a current to flow in the wires. In the infrared spectrometer this current will be proportional to the intensity of radiation falling on the thermopile.

2.4.4   MODE OF OPERATION

The following is a simplified description of the operation of a typical infrared machine, and figure 2.8 is a schematic representation of such a machine.

Light from the source (A) is split into two equal beams, one of which (B) passes through the sample (the *sample beam*), the other behaving as a *reference beam*; the function of such a *double-beam operation* is to measure the difference in intensities between the two beams at each wavelength.

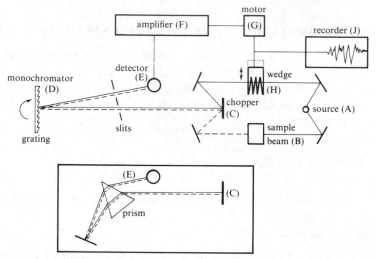

Figure 2.8   *Schematic layout of an infrared spectrometer. Main diagram— grating optics; insert—prism monochromator, which may replace the grating.*

The two beams are now reflected to a *chopper* (C), which consists of a rotating segmented mirror; as the chopper rotates (≈ 10 times per second) it causes the sample beam and the reference beam to be reflected *alternately* to the monochromator grating (D). As the grating slowly rotates, it sends individual frequencies to the detector thermopile (E), which converts the infrared (thermal) energy to electrical energy.

When a sample has absorbed light of a particular frequency, the detector will be receiving *alternately* from the chopper an intense beam (the reference beam) and a weak beam (the sample beam). This will in effect lead to a pulsating or *alternating* current flowing from the detector to the amplifier (F). (If the sample had not absorbed any light, the sample beam and the reference beam would have been of equal intensity, and the signal from the detector would have been *direct* current. The amplifier is designed only to amplify alternating current.)

The amplifier, which is now receiving this out-of-balance signal, is coupled to a small servo-motor (G), which drives an optical wedge (H) into the reference beam until eventually the detector receives light of equal intensity from sample and reference beams. This movement of the wedge (or attenuator) is in turn coupled to a pen recorder (J) so that movement of the wedge in and out of the reference beam shows as absorption bands on the printed spectrum.

Since this instrument balances out by optical means the differential between the two beams, it is a double-beam *optical-null* recording spectrometer.

For prism optics the principle is identical: the grating is replaced by a prism, and a rotating mirror effects the scanning of individual frequencies. The insert in figure 2.8 shows clearly how the prism assembly replaces the grating.

2.4.5 CALIBRATION OF THE FREQUENCY SCALE

Since the majority of instruments use preprinted recorder chart paper, the scanning of the frequencies and the driving of the recorder must be carefully adjusted so that accurate frequencies are traced onto the charts. Calibration can be carried out using the spectrum of polystyrene (or of indene); these spectra show many sharp bands whose frequencies are accurately known (see figures 2.9, 2.10 and 2.23). It is good practice to check the instrument calibration frequently, and even changes in humidity can cause inaccuracies because of the shrinking or stretching of the paper charts. All of the spectra in this book show the 1601 cm^{-1} peak of polystyrene, marked on as a check of frequency accuracy.

2.4.6 LINEAR-WAVENUMBER AND LINEAR-WAVELENGTH SCALES

The two polystyrene spectra in figures 2.9 and 2.10 have quite different appearances; the former is plotted on a linear-wavenumber abscissa scale, while the latter has a linear-wavelength abscissa. Most recent spectrometers use the linear-wavenumber presentation, and all other spectra in this book are so presented. The poor resolution obtained from NaCl prism optics, especially at high frequencies, makes it pointless to expand this region, which is very much compressed in figure 2.10 (prism-optics instrument) compared to figure 2.9 (grating-optics instrument).

With the advent of cheap grating instruments, together with a linear-wavenumber presentation, full advantage can now be taken of the remarkable resolution these machines give at high frequencies. (Note however that in figure 2.9, etc., there is a 2:1 gear change at 2000 cm^{-1}, which avoids overcrowding of the bands at low frequencies).

Students familiar with linear-wavelength spectra will only be very slightly disadvantaged when studying the spectra and charts in this book.

2.4.7 ABSORBANCE AND TRANSMITTANCE SCALES

The intensity of absorption bands in infrared spectra cannot easily be measured with the same accuracy as in ultraviolet spectra (see chapter 4). It is

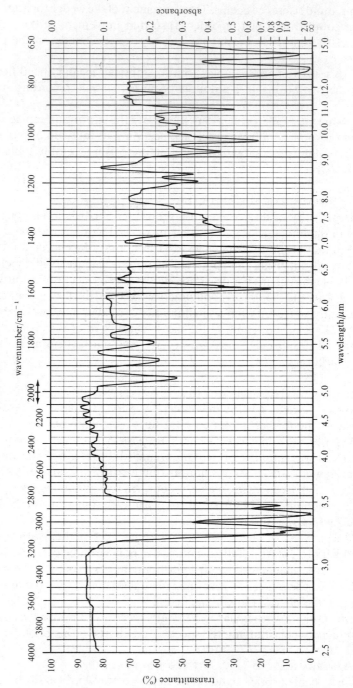

Figure 2.9 *Infrared spectrum of polystyrene. 0.05 mm film. Presented linear-in-wavenumber.*

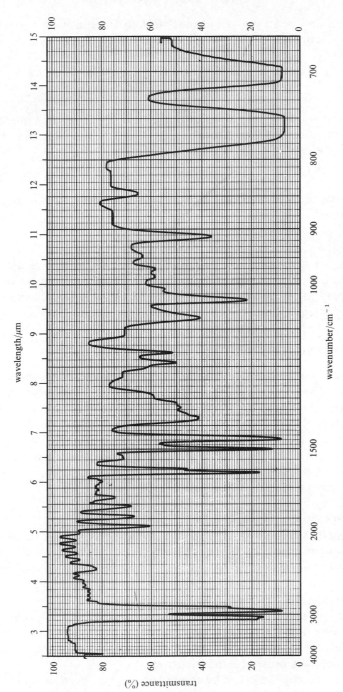

Figure 2.10 *Infrared spectrum of polystyrene. 0.05 mm film. Presented linear-in-wavelength.*

usually sufficient for an organic chemist to know that a band is of strong, medium, weak or variable intensity, indicated on charts, etc. as *s*, *m*, *w* or *v*, respectively.

The *absorbance* of a sample at a particular frequency is defined as

$$A = \log(I_0/I)$$

where I_0 and I are the intensities of the light before and after interaction with the sample, respectively. Absorbance is therefore a logarithmic ratio.

The *transmittance* of a sample is defined as

$$T = I/I_0$$

Transmittance therefore bears a reciprocal and logarithmic relationship to absorbance

$$A = \log(1/T)$$

The ordinate scale of figure 2.9 shows how these might be presented on a typical spectrum. For reasons that are largely instrumental, most infrared spectra record the intensities of bands as a linear function of *T*, and this has important consequences in quantitative work (see section 2S.1).

2.5   SAMPLING TECHNIQUES

A wide range of techniques is available for mounting the sample in the beam of the infrared spectrometer. These *sampling techniques* depend on whether the sample is a gas, a liquid or a solid. Intermolecular forces vary considerably in passing from solid to liquid to gas, and the infrared spectrum will normally display the effect of these differences in the form of frequency shifts or additional bands, etc. It is therefore most important *to record on a spectrum the sampling technique used.*

Various micro methods are dealt with in section 2S.4.

2.5.1   GASES

The gas sample is introduced into a *gas cell*, typically as shown in figure 2.11; infrared-transparent windows (for example NaCl) allow the cell to be mounted directly in the sample beam. In a modified form, the use of internal mirrors permits the beam to be reflected several times through the sample (*multi-pass* gas cells) to increase the sensitivity.

In the vapour phase, rotational changes in the molecule can occur freely, and these very low-frequency (low energy) processes can modulate the higher-energy vibrational bands; the vibrational bands are split, often with the production of considerable fine structure.

Few organic compounds can be examined as gases, even in heated cells, and gas sampling is not part of the average organic chemist's infrared repertoire.

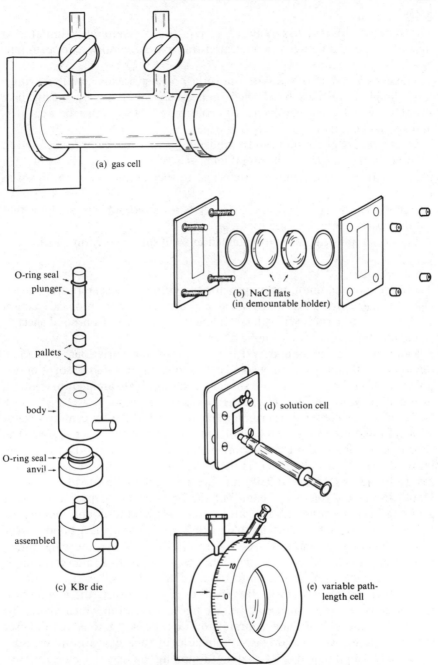

O-ring seal
plunger

pallets

body

O-ring seal
anvil

assembled

(a) gas cell

(b) NaCl flats
(in demountable holder)

(d) solution cell

(c) KBr die

(e) variable path-
length cell

Figure 2.11 *Equipment used in infrared sampling techniques: (a) gas cell; (b) sodium chloride flats (rock salt flats), and demountable holder; (c) KBr die (exploded view and assembled view); (d) solution or liquid cell being filled (e) variable path-length cell.*

2.5.2 LIQUIDS

The simplest infrared technique of all consists of sampling a liquid as a *thin film* squeezed between two infrared-transparent windows (for example, NaCl flats: see figure 2.11).

These *rock salt flats* must be optically polished (using jewellers' rouge for example) and must be cleaned immediately after use by rinsing in a suitable solvent such as toluene, chloroform, etc. They must be kept dry, and should be handled only by their edges.

The thickness of the film can be adjusted by varying the pressure used to squeeze the flats together; the film thickness is ≈ 0.1–0.3 mm. The assembled pair of flats and liquid film are mounted in the sample beam as shown in figure 2.11.

For spectra down to 250 cm^{-1}, CsI flats are used, and for samples that contain water, CaF_2 flats are used.

Liquid samples can also be examined in solution (see section 2.5.4).

2.5.3 SOLIDS

There are three common techniques for recording solid spectra: *KBr discs*, *mulls*, and *deposited films*. Solids can also be examined in solution (see section 2.5.4) but solution spectra may have different appearances from solid spectra, since intermolecular forces will be altered.

Many organic compounds exist as polymorphic variations, and these different crystalline forms may also lead to different infrared absorptions. If polymorphism is suspected, the substance should be examined in solution, where all polymorphic forms lose their differences.

KBr discs are prepared by grinding the sample (0.1–2.0 per cent by weight) with KBr and compressing the whole into a transparent wafer or disc. The KBr must be dry, and it is an advantage to carry out the grinding under an infrared lamp to avoid condensation of atmospheric moisture, which gives rise to broad absorption at 3500 cm^{-1}. This can be alleviated by having a blank disc in the reference beam, but the best remedy is prevention. The grinding is usually done with an agate mortar and pestel, although commercial ball mills are available; considerable work is needed to achieve good dispersion, and poorly ground mixtures lead to discs that scatter more light than they transmit. The particle size that must be achieved to avoid scattering is less than the wavelength of the infrared radiation, that is, less than 2 μm.

Compression to a cohesive disc requires high pressure, the commonest technique being to use a special die (see figure 2.11) from which air can be evacuated before hydraulic compression to about 10 tons load. Discs produced in this way are fairly easy to handle (by their edges!), and measure commonly 13 mm in diameter and 0.3 mm in thickness. Less expensive equipment can consist of the special die, used with a simple screw-jack; even a large steel nut with two bolts screwed in from opposite ends with the KBr between them can give satisfactory discs.

Mulls, or pastes, are prepared by grinding the sample with a drop of oil; the mull is then squeezed between transparent windows as for liquid samples. The mulling agent should ideally be infrared-transparent, but this is never true, and the spectrum produced always shows the absorptions of the mulling agent superimposed on that of the sample.

Liquid paraffin (Nujol), whose infrared spectrum is shown in figure 2.1, is transparent over a wide range, and *Nujol mulls* are by far the most widely utilised. The absorptions of the Nujol effectively blank out the regions of C—H stretch and C—H bend, but these regions can be studied by preparing a mull with a complementary agent containing no C—H bonds: hexachlorobutadiene and chlorofluorocarbon oils are used for this purpose.

The regions of absorption of these three mulling agents are shown in figure 2.12.

Solid films can be deposited onto NaCl flats by allowing a solution in a volatile solvent to evaporate drop by drop on the surface of the flat. Polymers and various waxy or fatty materials often give excellent spectra in this way, but in many other cases the film is too sharply crystalline and therefore opaque.

2.5.4 SOLUTIONS

The sample can be dissolved in a solvent such as carbon tetrachloride, carbon disulphide or chloroform, and the spectrum of this solution recorded. The solution (usually 1–5 per cent) is placed in a *solution cell* consisting of transparent windows with a spacer between them of known thickness; the spacer is commonly of lead or polytetrafluoroethylene, and its thickness determines the path length of the cell–usually 0.1–1.0 mm. A second cell containing pure solvent is placed in the reference beam, so that solvent absorptions are cancelled out and the spectrum recorded is that of solute alone.

Although solvent cancellation is in principle possible throughout the entire range, in those regions where the solvent has strong absorption bands too little light passes to the detector (in either sample or reference beam) for the detector to respond sensitively. For this reason two spectra should be run using solvents whose transparent regions are complementary. Figure 2.12 shows where the common solvents have strong absorption bands; note that carbon tetrachloride and carbon disulphide are almost perfectly complementary, whereas chloroform is less suitable and is usually only called upon because of its superior solvent powers.

For really careful work it should be appreciated that using two cells of the same path length will lead to over-compensation of solvent absorptions in the reference beam; as an example, with 5 per cent solute and 95 per cent solvent in the sample beam, exact compensation would be achieved when the reference cell has a path length only 95 per cent of that of the sample cell. When this is important (as in quantitative measurements, see section 2S.1)

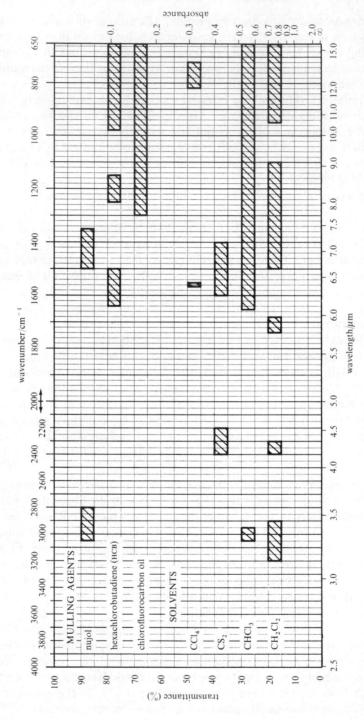

Figure 2.12 *Infrared absorption regions for common mulling agents and solvents. (Not all of these absorption regions contain strong bands).*

a *variable path-length cell* can be used whose path length can be accurately adjusted in deference to the concentration of the solution (see figure 2.11).

Solution spectra, with their absence of complicating features introduced by intermolecular forces and polymorphism, in general make the exact comparison of infrared data from different sources more valid. The solvent, however, must always be specified on the spectrum, since band frequencies do shift on changing to solvents of differing polarity, etc. This is particularly true of C=O bands and hydrogen-bonded systems (see section 2.3.2), and for this reason routine organic solution spectra are invariably recorded using carbon tetrachloride, carbon disulphide, chloroform, and occasionally dichloromethane, as solvents.

2.6 APPLICATIONS OF INFRARED SPECTROSCOPY— IDENTITY BY FINGERPRINTING

Infrared spectra contain many absorptions associated with the complex interacting vibrating systems in the molecule, and this pattern of vibrations, since it is uniquely characteristic of each molecule, gives rise to a uniquely characteristic set of absorption bands in the spectrum.

This band pattern serves as a *fingerprint* of the molecule; the region that contains a particularly large number of unassigned vibrations (and is most valuable in this respect) is roughly from 900–1400 cm^{-1}, and this general area is often called the *fingerprint region*.

To identify an unknown compound, one need only compare its infrared spectrum with a set of standard spectra recorded under identical conditions.

Substances that give the same infrared spectra are identical.

This proof of identity is far more characteristic than the comparison of any other physical property.

Having stated the principle, a few cautionary words should be added. For two spectra to be really identical would involve recording the spectra on the same machine under identical conditions of sampling, scan speed, slit widths, etc. Where this condition does not apply, some discretion must be allowed, but in general the greater the number of peaks in the fingerprint region the more reliable the proof of identity.

Many industrial companies have digitalised the information carried in the infrared spectra of standard compounds and stored this in computer memory banks. The spectrum from an unknown compound can be fed into the same digital storage; the computer is instructed to compare it with the standards and to seek out that compound which has identical infrared absorptions to the unknown. The principal limitation here is in the number of standard compounds whose spectra can be stored in even a relatively large computer; no computer has yet been built which is large enough to store information from the known two million organic compounds, far less accommodate the 100 000 new compounds produced annually.

Small changes in large molecules may produce very little change in the spectrum. For example, the infrared spectrum of a C_{20} straight-chain alkane is quite indistinguishable from that of its next higher straight-chain homologue. For a distinction of this magnitude mass spectroscopy would be the method of choice (see chapter 5).

2.7   APPLICATIONS OF INFRARED SPECTROSCOPY— IDENTIFICATION OF FUNCTIONAL GROUPS

By examining a large number of compounds known to contain a functional group, we can establish which infrared absorptions are associated with that functional group; we can also assess the range of frequencies within which each absorption should appear. Exactly this kind of information is set out in the *correlation charts* which follow (pp. 40–51).

Now, working in the converse, if we have an unknown compound whose functional groups we wish to identify, we can examine its infrared spectrum and use the correlation data to deduce which are the functional groups present.

It is impossible to arrive at a wholly systematic method for dealing with an infrared spectrum. All other evidence should be assessed simultaneously, be it chemical, physical or spectroscopic; even the known history of the compound can be revealing. It is *not* possible to identify a compound merely by interpreting its infrared spectrum from correlation data.

We shall discuss in detail the strengths and weaknesses of the method in the following pages but it is useful now to make a few clear statements on the general principles involved.

(i) Most weight can be placed on the absorptions above 1400 cm^{-1} and below 900 cm^{-1}. (The fingerprint region, 900–1400 cm^{-1}, contains many unassigned absorptions.)

(ii) *Group frequencies* are more valuable than single absorption bands. In other words, a functional group that gives rise to *many* characteristic absorptions can usually be identified more definitely than a function that gives rise to only one characteristic absorption. (Thus ketones (C=O *str*) are less easily identified than esters (C=O *str* and C—O *str*); esters are less easily identified than amides (C=O *str*, N—H *str*, N—H *def*, etc.)

(iii) The absence of a characteristic absorption may be more illuminating than its presence. (Consider the relative implications of the presence or absence of a C=O *str* absorption.)

(iv) Multifunctional compounds will show the separate absorptions of the individual functional groups, unless these interact. (Examples of *interacting functional types* are β-diketones, aliphatic amino acids, γ-hydroxy acids, etc.)

(v) The frequencies shown graphically on the correlation charts do not take account of any exceptional features in specific molecules; important circumstances, which might in this way lead to frequency shifts outside the quoted ranges, are discussed below.

(vi) *Graphically presented correlation charts are invariably sufficiently accurate for functional group identification*; coupled with this, frequencies of bands cannot easily be measured more accurately than ± 5 cm^{-1} on low-cost routine instruments (with lesser accuracy at higher frequencies).

More accurate tabular data are to be found in the specialist texts of Bellamy, Cross and van der Maas, and such tables should be used for studying the restricted frequency ranges of more narrowly specific classes of organic molecule. The discussions of group frequencies that follow contain some more detailed frequency data (see sections 2.8–2.13).

A reasonable initial plan of attack on the infrared spectrum of a totally unknown compound would be to spend a few minutes searching out the most commonly successful correlations.

The carbon skeleton should be tackled first (see section 2.8): look for evidence of alkane, alkene, alkyne and aromatic residues (using C—H *str*, C—H *def* and the various carbon–carbon bond stretching frequencies). Evidence from the n.m.r. spectrum is of great complementary value.

Look for C=O *str*; if present, it may be associated with C—H *str* in aldehydes, N—H *str* in amides, C—O *str* in esters, etc.

Look for O—H *str* or N—H *str*.

Look for C≡N *str*.

In sulphur compounds look for S—H *str*, S=O *str*, and —SO$_2$— *str*.

In phosphorus compounds look for P=O *str*.

Chemical and/or mass spectrometric evidence for the presence of nitrogen, sulphur, halogens or phosphorus, etc., is essential: infrared evidence for these is self-sufficient *only* in exceptional cases.

The correlation charts are set out in the order indicated by the above approach to an infrared identification of functional groups. Thus 'aromatic carbon skeletons' precede alkane, since the former are frequently easier to confirm. The charts themselves contain commentary, which will frequently suffice to make an assignment, but additional discussion of each class is given after the charts.

The frequency ranges shown on the charts apply to spectra recorded on liquid films or KBr discs, or Nujol mulls, etc., since most routine spectra will be recorded thus. Where dilute solution spectra produce substantial shifts this is indicated on the charts and in the discussion following.

CHART 1 contains assignments that will identify the nature of *the carbon skeleton*, identifying (i) aromatic, (ii) alkane, (iii) alkene and (iv) alkyne residues. (See also section 2.8.)

CHART 2 is concerned with *carbonyl compounds*, including aldehydes, ketones, enols and all carboxylic acid derivatives. (See also section 2.9.)

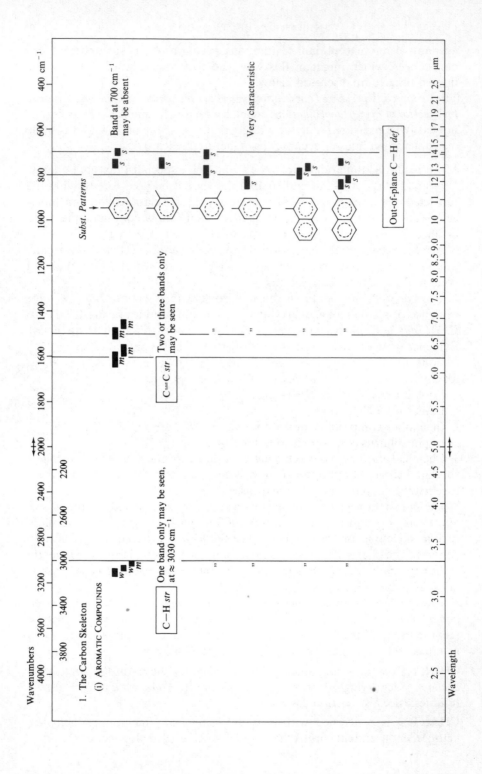

Wavenumbers

1. The Carbon Skeleton

(i) AROMATIC COMPOUNDS

C—H *str*
One band only may be seen,
at ≈ 3030 cm⁻¹

C═C *str*
Two or three bands only
may be seen

Subst. Patterns →

Band at 700 cm⁻¹
may be absent

Very characteristic

Out-of-plane C—H *def*

Wavelength

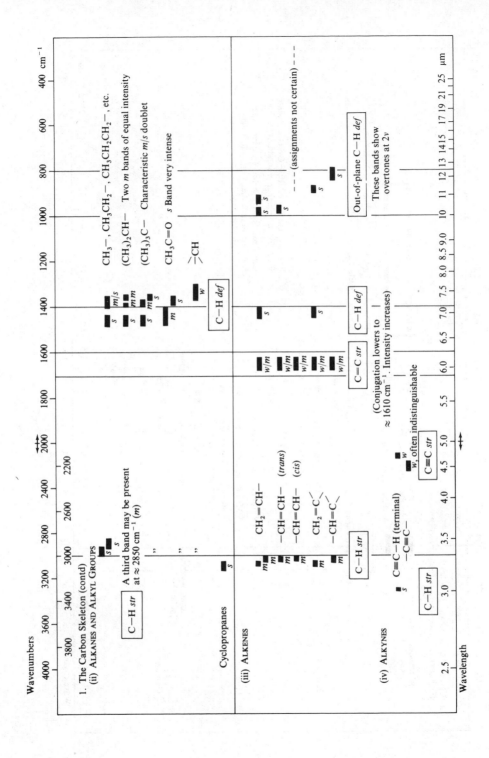

Wavenumbers

1. The Carbon Skeleton (contd)
(ii) ALKANES AND ALKYL GROUPS

CH_3-, CH_3CH_2-, $CH_3CH_2CH_2-$, etc.
$(CH_3)_2CH-$ Two m bands of equal intensity
$(CH_3)_3C-$ Characteristic m/s doublet
$CH_3C=O$ s Band very intense
$\geq CH$

C—H str A third band may be present at ≈ 2850 cm^{-1} (m)

C—H def

Cyclopropanes

(iii) ALKENES

$CH_2=CH-$
$-CH=CH-$ (trans)
$-CH=CH-$ (cis)
$CH_2=C\diagdown$
$-CH=C\diagup$

C—H str

C=C str (Conjugation lowers to ≈ 1610 cm^{-1}. Intensity increases)

C—H def

Out-of-plane C—H def These bands show overtones at 2ν
- - - (assignments not certain) - - -

(iv) ALKYNES

$C≡C—H$ (terminal)
$-C≡C-$

C—H str

$C≡C$ str w, often indistinguishable

Wavelength

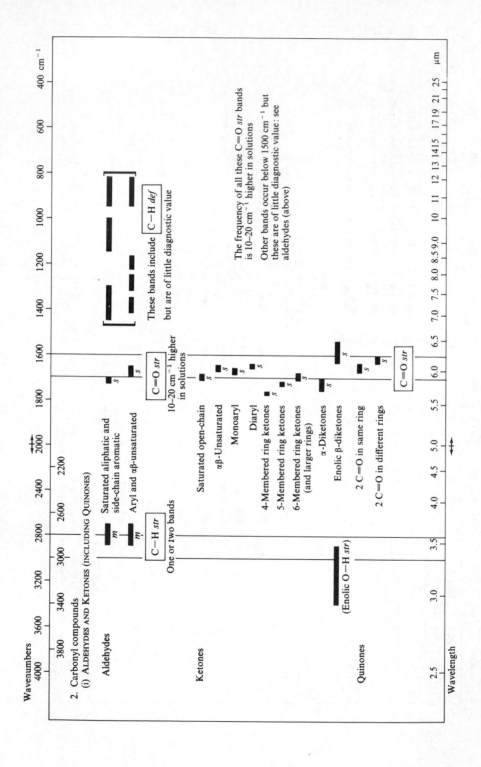

2. Carbonyl compounds
(i) ALDEHYDES AND KETONES (INCLUDING QUINONES)

Aldehydes

Saturated aliphatic and side-chain aromatic

Aryl and αβ-unsaturated

C—H str
One or two bands

C=O str
10–20 cm⁻¹ higher in solutions

Ketones

Saturated open-chain

αβ-Unsaturated

Monoaryl

Diaryl

4-Membered ring ketones

5-Membered ring ketones

6-Membered ring ketones (and larger rings)

α-Diketones

Enolic β-diketones

2 C=O in same ring

2 C=O in different rings

C=O str

These bands include C—H def but are of little diagnostic value

The frequency of all these C=O str bands is 10–20 cm⁻¹ higher in solutions

Other bands occur below 1500 cm⁻¹ but these are of little diagnostic value: see aldehydes (above)

Quinones

(Enolic O—H str)

Wavenumbers
3800   3600   3400   3200   3000   2800   2600   2400   2200   2000   1800   1600   1400   1200   1000   800   600   400 cm⁻¹

Wavelength
2.5   3.0   3.5   4.0   4.5   5.0   5.5   6.0   6.5   7.0 7.5 8.0 8.5 9.0 10 11 12 13 14 15 17 19 21 25   μm

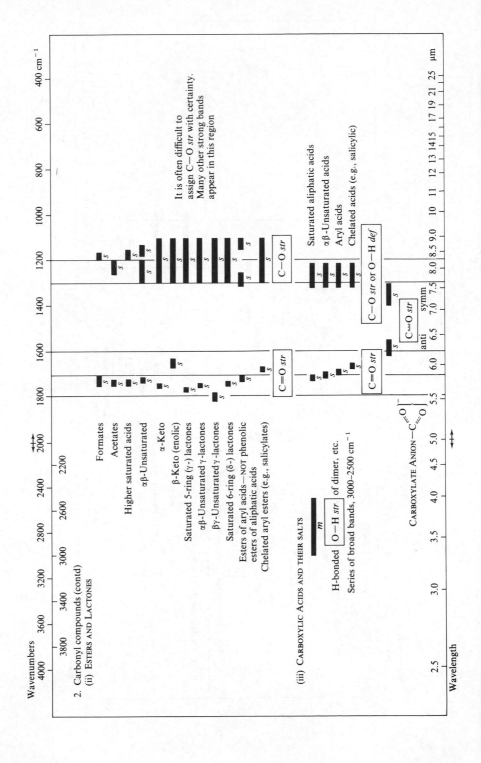

2. Carbonyl compounds (contd)
(ii) ESTERS AND LACTONES

Formates
Acetates
Higher saturated acids
αβ-Unsaturated

α-Keto

β-Keto (enolic)
Saturated 5-ring (γ-) lactones
αβ-Unsaturated γ-lactones
βγ-Unsaturated γ-lactones
Saturated 6-ring (δ-) lactones
Esters of aryl acids—NOT phenolic
esters of aliphatic acids
Chelated aryl esters (e.g., salicylates)

C=O str

C—O str

It is often difficult to
assign C—O str with certainty.
Many other strong bands
appear in this region

(iii) CARBOXYLIC ACIDS AND THEIR SALTS

H-bonded O—H str of dimer, etc.

Series of broad bands, 3000–2500 cm⁻¹

Saturated aliphatic acids
αβ-Unsaturated acids
Aryl acids
Chelated acids (e.g., salicylic)

C=O str

C—O str or O—H def

CARBOXYLATE ANION —C(=O)(—O)⁻

C=O str

anti   symm

Wavenumbers

Wavelength

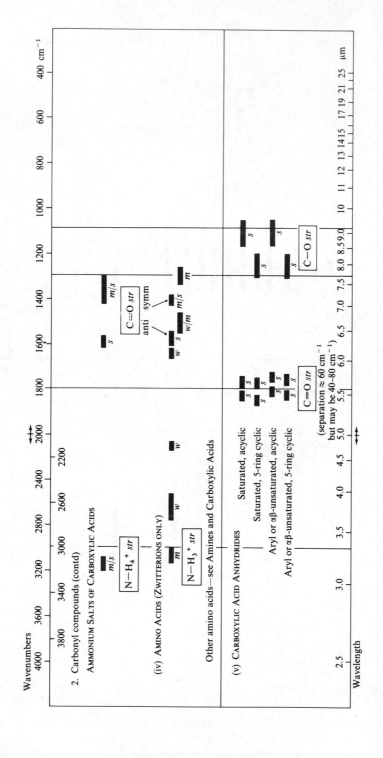

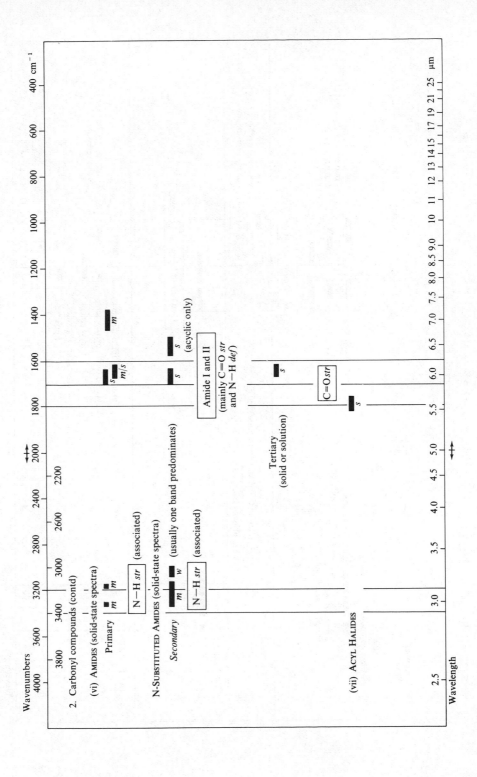

Wavenumbers

2. Carbonyl compounds (contd)

(vi) AMIDES (solid-state spectra)

Primary

N—H *str* (associated)

N-SUBSTITUTED AMIDES (solid-state spectra)

Secondary (usually one band predominates)

N—H *str* (associated)

Amide I and II
(mainly C=O *str*
and N—H *def*)

(acyclic only)

Tertiary
(solid or solution)

C=O *str*

(vii) ACYL HALIDES

Wavelength

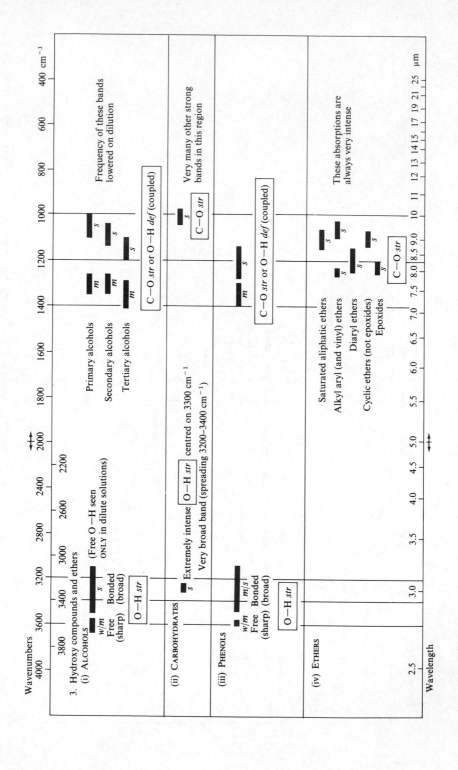

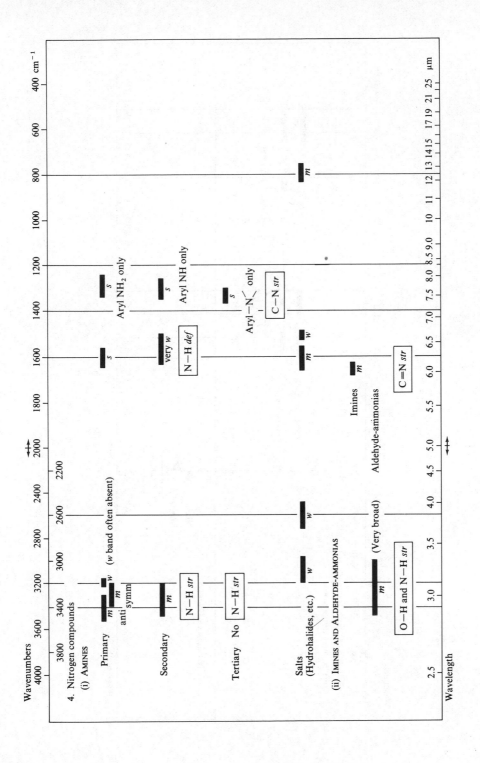

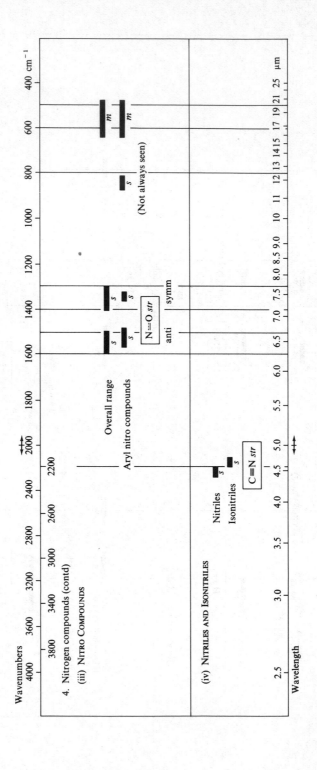

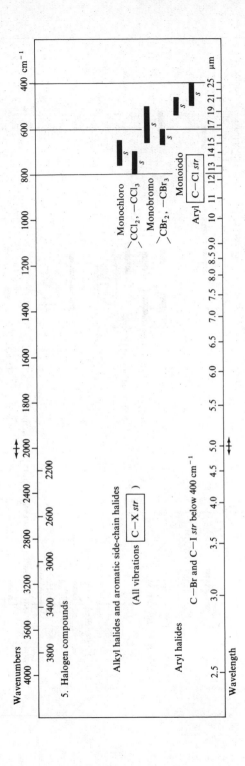

5. Halogen compounds

Wavenumbers

4000 3800 3600 3400 3200 3000 2800 2600 2400 2200 2000 1800 1600 1400 1200 1000 800 600 400 cm^{-1}

Alkyl halides and aromatic side-chain halides

(All vibrations $\boxed{\text{C—X } str}$)

Monochloro

$\diagdown$CCl$_2$, —CCl$_3$

Monobromo

$\diagdown$CBr$_2$, —CBr$_3$

Monoiodo

Aryl halides

Aryl $\boxed{\text{C—Cl } str}$

C—Br and C—I str below 400 cm^{-1}

2.5 3.0 3.5 4.0 4.5 5.0 5.5 6.0 6.5 7.0 7.5 8.0 8.5 9.0 10 11 12 13 1415 17 19 21 25 μm

Wavelength

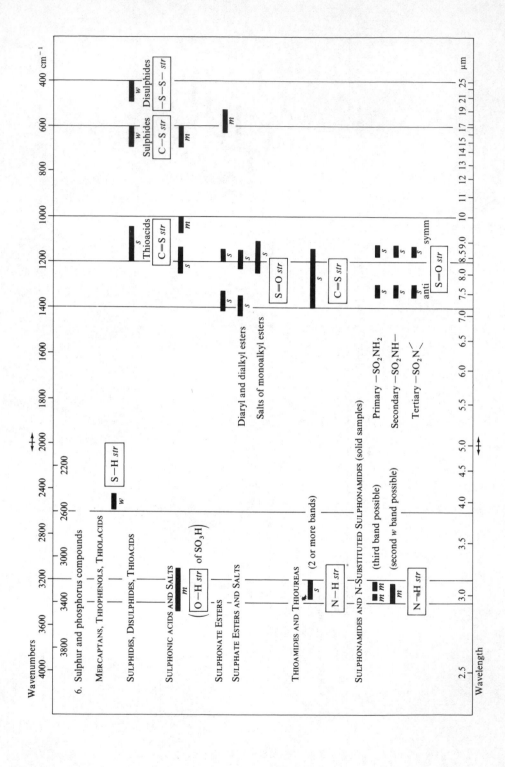

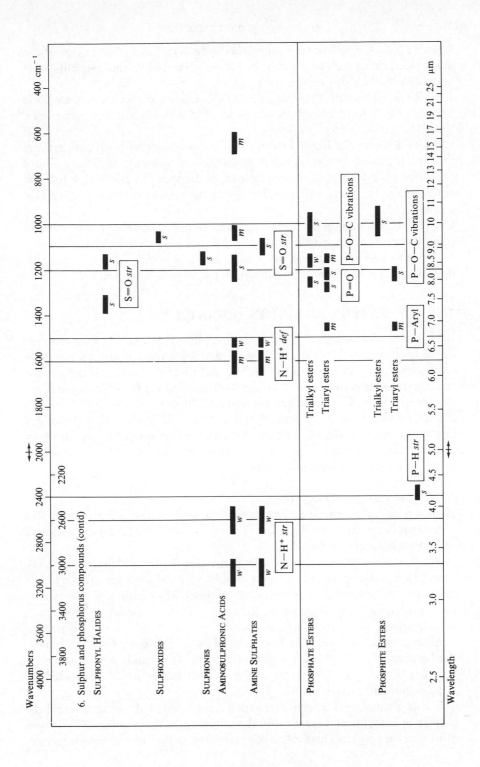

6. Sulphur and phosphorus compounds (contd)

CHART 3 deals with *hydroxy compounds and ethers*, but excluding carboxylic acids (see previous chart); alcohols, phenols, epoxides, and carbohydrates appear here. (See also section 2.10.)

CHART 4 begins with the assignments of *nitrogen compounds* that are basic (amines, etc.); nitro compounds, nitriles and isonitriles also appear here. (See also section 2.11.)

CHART 5 shows the limited number of carbon–halogen bond assignments that routine infrared spectra afford in *halogen compounds*; acyl halides have the more valuable C=O *str* correlation, and appear in chart 2: sulphonyl halides have the more valuable S=O *str* and appear in chart 6. (See also section 2.12.)

CHART 6 covers *sulphur and phosphorus compounds*, which may contain other heteroatoms (nitrogen and halogen). (See also section 2.13.)

Abbreviations: s = strong, m = medium, w = weak and v = variable intensity, respectively: anti = antisymmetric, symm = symmetric.

2.8   THE CARBON SKELETON (CHART 1)

In deducing the nature of carbon residues in an organic molecule by infrared spectroscopy, it is useful to note that (a) aromatic groups are most easily detected from C∺C *str* and C—H *def* absorptions, (b) alkene groups are easily detected from C=C *str* absorptions, unless aromatic residues are also present, (c) alkane residues are detected from C—H *str* and C—H *def* absorptions (but n.m.r. in general is more specific in detecting particular groupings such as Me, Et, Pr^n, Pr^i, Bu^t) and (d) terminal alkynes are very easily detected from C≡C *str* and C—H *str* absorptions, but nonterminal alkynes may cause extreme difficulty.

2.8.1   AROMATICS (CHART 1(i))

Four regions of the spectrum are associated with the confidently assignable vibrations of aromatic residues (C—H *str*, C∺C *str*, C—H *def* and a group of overtone-combination bands).

The positions of C—H *str* absorptions are shown on the charts: these weak absorptions may appear merely as a shoulder on the stronger alkane C—H *str* bands, and indeed may be swamped by them. The distinction can usually be clearly made that aromatic and alkene C—H *str* is just *above* 3000 cm^{-1}, while alkane C—H *str* is just below.

Most aromatic compounds show three of the four possible C∺C *str* bands: the band at 1450 cm^{-1} is often absent, and the two bands near 1600 cm^{-1} occasionally coalesce, so that two, three or four bands must be expected for this assignment.

Out-of-plane C—H deformations are strongly coupled vibrations, and the pattern of absorption is reasonably characteristic of the number of hydrogen atoms on the ring (and hence also characteristic of the *substitution pattern* on the ring). The most reliable is the strong band above 800 cm^{-1} for *p*-substituted benzenes. The out-of-plane C—H *def* bands are strong and broad, being typically 30–50 cm^{-1} wide at half height, and are therefore easily identified on the spectrum. Because of the ambiguity in band positions, n.m.r. evidence for substitution pattern should always accompany infrared evidence. Chart 1 shows the C—H *def* absorptions for benzene and naphthalene rings: corresponding data for heterocyclic rings, etc., is given in Bellamy (see Further Reading).

Bands corresponding to aromatic C—H *str*, C==C *str* and out-of-plane C—H *def* are clearly seen in figures 2.3, 2.4, 2.9 and 2.10. The aromatic C—H *str* bands can *not* be assigned in figures 2.7 and 2.15.

The bands shown in figures 2.9 and 2.10 (polystyrene film) between 1650 and 2000 cm^{-1} are overtone and combination bands mainly of the aromatic C—H *def* absorptions: the nature of such bands is discussed in section 2.2.2. They appear in all spectra of aromatic compounds, but are usually too weak to be clearly interpreted unless large sample sizes are used. Since they are related to the out-of-plane C—H *def* vibrations, they too show a dependence on ring substitution; the band positions do not move on changing substitution; rather do their relative intensities alter. They are now seldom used for structural deductions where n.m.r. evidence is available.

2.8.2 ALKANES AND ALKYL GROUPS (CHART 1(ii))

Both C—H *str* and C—H *def* absorptions in saturated aliphatic groups are strong or medium absorptions, and there is seldom difficulty in assigning these bands on the spectrum.

By far the commonest appearance of C—H *str* absorptions is two strong bands just below 3000 cm^{-1} (of which the higher frequency is antisymmetric stretch). See figures 2.1, 2.2, 2.6, 2.9 and 2.10: these last two figures illustrate nicely that aliphatic C—H *str* lies just below 3000 cm^{-1}, with aromatic C—H *str* just above. In spectra recorded on very high-resolution instruments, simple alkanes show four C—H *str* bands corresponding to CH_3 and CH_2 antisymmetric C—H *str* (high-frequency pair) and the corresponding symmetric C—H *str* (pair at lower frequency): this degree of resolution is exceptional in compounds other than simple alkanes.

The pattern of bands corresponding to C—H *def* may be characteristic of the alkyl groups present, *provided* no strong electrical influence is close by in the molecule. In particular the antisymmetric CH_3 *def* around 1390 cm^{-1} is split in $(CH_3)_2C=$ and $(CH_3)_3C—$ groups, and such groups can be detected with reasonable certainty from the infrared spectrum (see figure 2.14): otherwise, the commonest appearance of the C—H *def* bands is as shown in figures 2.1 and 2.6, which gives no clue to the alkyl grouping present. In general, n.m.r. evidence is preferred.

Other bands in the infrared spectra of alkanes and alkyl groups are of doubtful origin; the bands from 800–1300 cm^{-1} should be considered as C—C *str* or C—H *def*, and thereafter disregarded.

2.8.3   ALKENES (CHART 1(iii))

Most of the useful assignments are shown on the correlation chart for alkenes.

The similarity to aromatic absorptions is immediately apparent, and it is often difficult to detect an alkene group *within an aromatic molecule by* infrared spectroscopy: conjugation of the double bond with phenyl lowers C=C *str* to around 1630 cm^{-1}.

Out-of-plane C—H *def* bands are easily seen, and are usually strong enough for their overtone bands to be substantial: since $2v$ lies in the region 1800–2000 cm^{-1} where few other absorptions appear, the overtone band may be prominent. *Trans*-alkenes can often be distinguished from the *cis* isomer by the former's C—H *def* band around 970 cm^{-1}; the corresponding band for *cis* isomers is often around 700 cm^{-1}, but it is not so certainly identified as that of the *trans* isomer.

Tetrasubstituted alkene groups such as $R_2C{=}CR_2$ show either no C=C *str* band or, at best, a weak band: they do show a Raman band, however (see section 2S.3).

The infrared spectrum of limonene (figure 2.13) shows alkene C—H *str* and C=C *str*: the C—H *def* absorption around 900 cm^{-1} shows an overtone band around 1800 cm^{-1}.

Cumulative double bonds, as in allenes, give rise to strong absorptions around 2000 cm^{-1}: these are thought to be the antisymmetric C=C *str*, etc. and a few of the more common examples are: C=C=C (allenes), 1950 cm^{-1}: O=C=O, 2350 cm^{-1}: C=C=O (ketenes), 2150 cm^{-1}: —N=C=O (isocyanates), 2250 cm^{-1}: —N=C=S (isothiocyanates), 2100 cm^{-1}: —N$_3$ (azides), 2140 cm^{-1}.

2.8.4   ALKYNES (CHART 1(iv))

Terminal alkynes are easily detected by the co-presence of C—H *str* (strong band near 3300 cm^{-1}) and the weaker C≡C *str* band near 2200 cm^{-1}: see figure 2.2. Nonterminal alkynes are one of the most difficult classes to detect spectroscopically, since they lack the valuable alkyne C—H *str* absorption, and C≡C *str* is extremely weak: Raman spectroscopy is necessary to help identify this group (see section 2S.3).

2.9   CARBONYL COMPOUNDS (CHART 2)

The stretching absorption of the carbonyl group has been the subject of more study than any other absorption in the infrared; consequently our knowledge of the factors that give rise to frequency shifts is extensive. These factors have been discussed in section 2.3.

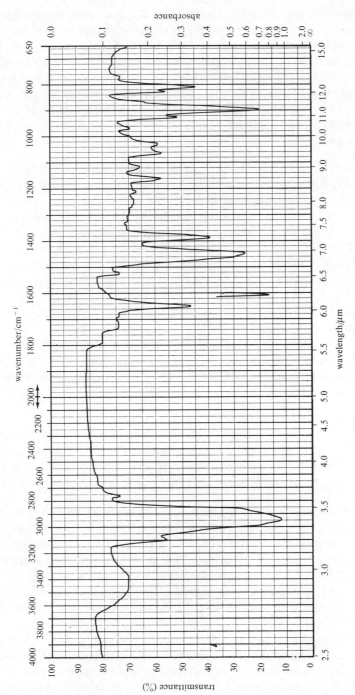

Figure 2.13 *Infrared spectrum of limonene. Liquid film.*

While the presence of C=O *str* absorption is almost always easily detected on the spectrum, identification of the carbonyl-containing function is not always feasible simply by noting the C=O *str* frequency. As ever, the functions that are easiest to detect are those that give rise to the greatest number of characteristic absorptions. Chemical and/or n.m.r. evidence is particularly helpful in the case of aldehydes, quinones, esters, carboxylic acids and amides.

The C=O *str* band is always strong, and consequently its overtone at $2v$ is often visible around 3400 cm^{-1}: this overtone band can be clearly seen in figures 2.4 and 2.14.

2.9.1   ALDEHYDES AND KETONES (INCLUDING QUINONES) (CHART 2(i))
Aldehydes are usually distinguishable from ketones, etc. by their C—H *str* absorptions: in both classes the absorptions below 1500 cm^{-1} are of little diagnostic value, one possible exception being the case of methyl ketones (see figure 2.4), which give rise to a very strong characteristic absorption just below 1400 cm^{-1} (noted on chart 1(ii)). Enols are easily identified by the broad H-bonded O—H *str* absorption, and by the very low C=O *str* frequency—as low as 1580 cm^{-1} in acetylacetone (which is about 85 per cent enolic).

It is important to note the narrow limits of C=O *str* frequencies for these compounds, and an industrious attempt must be made to relate this frequency to the detailed environment of the carbonyl group. A good example would be the cyclic ketones, where ring size can be gauged accurately from the position of the C=O *str* band. Such assignments must be in accord with deductions concerning the carbon skeleton.

2.9.2   ESTERS AND LACTONES (CHART 2(ii))
In addition to accurately known C=O *str* frequencies, the C—O *str* band(s) for esters and lactones can be highly informative. Thus most alkanoate esters show one C—O *str* band, while aryl and αβ-unsaturated carboxylate esters show two. (See bands at 1120 and 1280 cm^{-1} in figure 2.14).

2.9.3   CARBOXYLIC ACIDS AND THEIR SALTS (CHART 2(iii))
Carboxyl groups are one of the easiest functions to detect by infrared spectroscopy because of the co-presence of C=O *str* with the exceedingly broad O—H *str* absorption centred around 3000 cm^{-1}. The infrared spectrum of benzoic acid (figure 2.7) is typical. The broad O—H *str* band corresponds to the dimer structure discussed in section 2.3: O—H *def* in the dimer gives rise to a characteristic band near 950 cm^{-1}, seen in the benzoic acid spectrum in figure 2.7.

Carboxylic acid salts are most accurately represented with the resonance-stabilised carboxylate anion, which contains two identical C⫤O bonds, and therefore gives antisymmetric and symmetric stretching absorptions. Very reliable proof of identity involves converting the salt to the free acid, when the true C=O *str* absorption appears at higher frequency than C⫤O *str*.

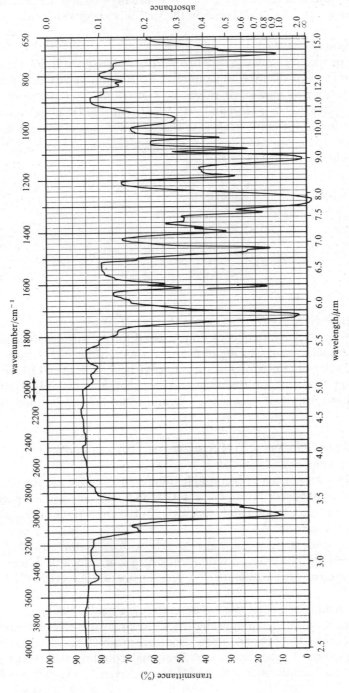

Figure 2.14 *Infrared spectrum of isoamyl benzoate. Liquid film.*

Many carboxylate salts contain water of crystallisation, which produces broad absorption around 3400 cm^{-1}. It is usually possible nevertheless to separate this from N—H *str* bands in the specific case of ammonium salts.

2.9.4   AMINO ACIDS (CHART 2(iv))

Amino acids with the amino group directly attached to an aryl ring exist with the *free* amino and carboxyl functions: all other amino acids (including all naturally occurring members) exists as zwitterions, and therefore show absorptions due to —$\overset{\oplus}{\text{N}}$—H *str* and —$CO_2{}^{\ominus}$ *str*. (Infrared studies historically provided some of the best evidence for the very existence of zwitterions.) On acidification the protonated amino acid then gives rise to true C=O *str* at higher frequency than C⁚=O *str*.

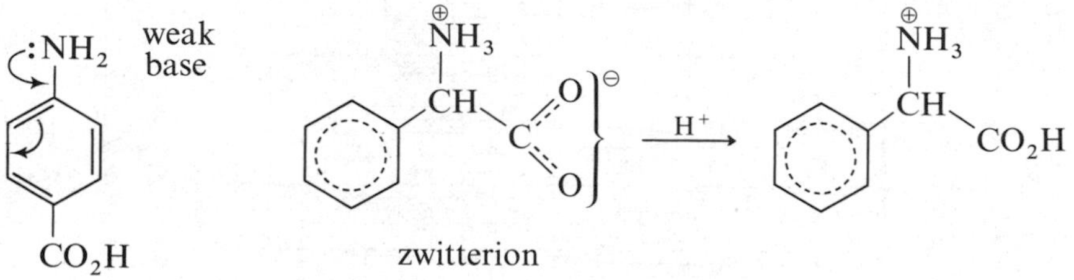

zwitterion

2.9.5   CARBOXYLIC ACID ANHYDRIDES (CHART 2(v))

The characteristic high-frequency double band for anhydrides makes them very easy to detect: the splitting of the band is due to Fermi resonance (see section 2.3). A common contaminant is the free acid, with O—H *str* absorption around 3000 cm^{-1}. Interestingly, imides also give rise to two coupled C=O vibrations, their frequencies being 1680–1710 cm^{-1} and 1730–1790 cm^{-1}, respectively.

2.9.6   AMIDES (PRIMARY AND N-SUBSTITUTED) (CHART 2(vi))

No liquid amides exist.

The infrared spectrum of benzamide (figure 2.15) shows the very characteristic pair of bands for N—H *str* near 3200 and 3400 cm^{-1}: the separation between these bands in solid-state spectra (120–180 cm^{-1}) is usually sufficient to identify primary amides.

The two bands are not associated simply with antisymmetric and symmetric N—H *str*, but arise from different degrees of association in amides: indeed a third band (near 3300 cm^{-1}) is present, but not observed in KBr-disc sampling.

Amides readily self-condense, as do carboxylic acids, but in amides the situation is additionally complicated by (a) the considerable double-bond character of the CO—N bond due to mesomerism which may lead to *cis* and *trans* forms, and (b) the degree of vibrational coupling that takes place (N—H *str* may couple both with N—H *str* and even N—H *def*: N—H *def*

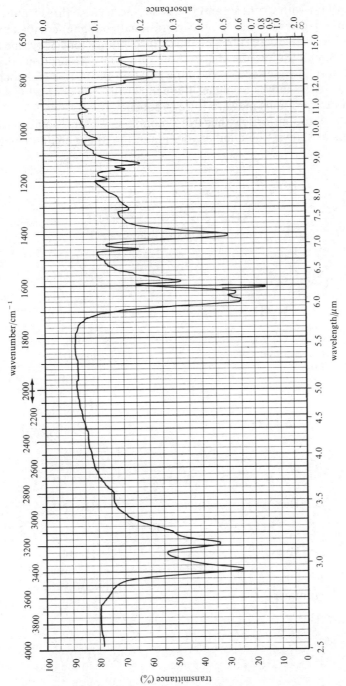

2.15 *Infrared spectrum of benzamide. KBr disc.*

couples strongly with C=O *str* and C—N *str*, etc.). These three factors (self-association, *cis–trans* isomerism, and vibrational coupling) completely rule out a simple reductive description of amide absorptions: for clarity, and at the risk of oversimplification, the charts show only those absorptions that are important in solid-state spectra. In dilute solution the degree of association changes, N—H *str* moves to higher frequency while the coupled C=O *str* and N—H *def* vibrations move to lower frequency.

N—H *str* vibrations arise either from monomer, dimer, chains of dimers or polymers as shown: the number of units in the dimer chain or polymer

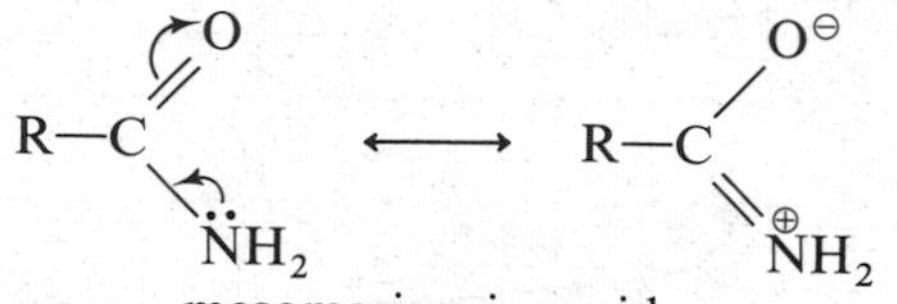

mesomerism in amides

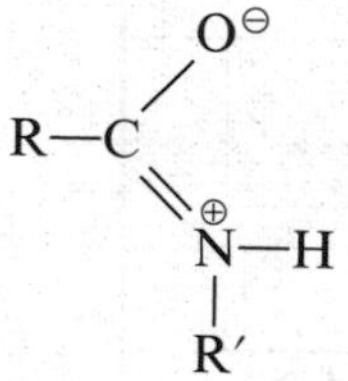

cis monomer 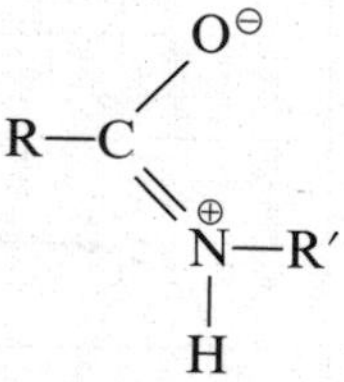　　　　　　　　*trans* monomer

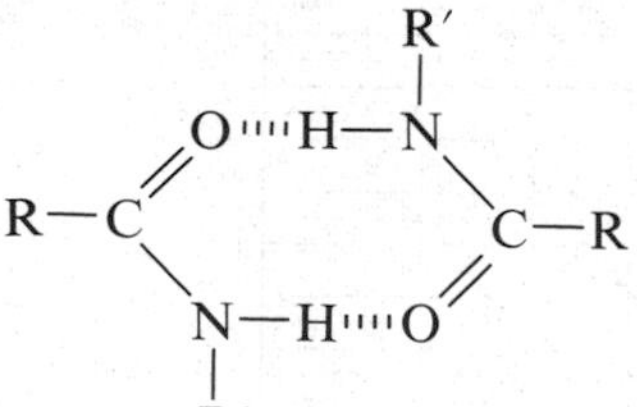

cis dimer

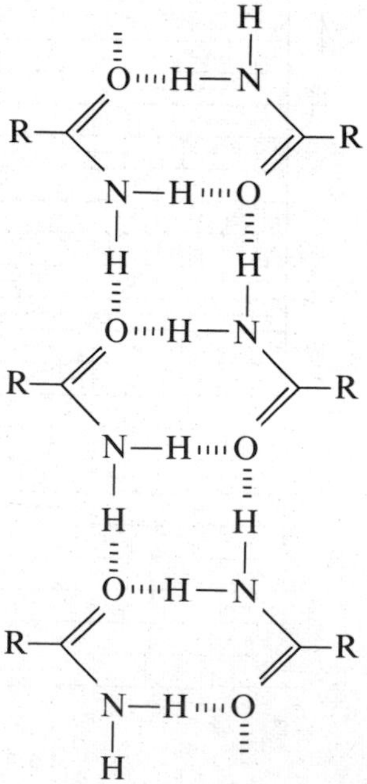

chains of dimers

$$\ce{...H} \quad \ce{H} \quad \ce{R} \quad \ce{O}...\ce{H} \quad \ce{H} \quad \ce{R} \quad \ce{O}...$$

polymer

determines the N—H *str* frequencies, which therefore shift with concentration changes. Primary amides in very dilute solution give N—H *str* near 3400 and 3500 cm^{-1}: secondary near 3450 cm^{-1}.

C=O *str* and N—H *def*, being coupled vibrations, give rise to bands that are best named simply as amide I and II bands respectively: these can be seen in figure 2.15 near 1650 cm^{-1}, and this double band is also highly characteristic of amides. The higher-frequency band is predominantly C=O *str*, and the lower is predominantly N—H *def*. Other amide bands of mixed assignment are named amide III, IV, V and VI.

2.9.7 ACYL HALIDES (CHART 2(vii))
The high-frequency C=O *str* band for acyl halides is reasonably characteristic, but chemical or mass spectral evidence is needed to detect and identify the halogen present.

2.10 HYDROXY COMPOUNDS AND ETHERS (CHART 3)

Hydrogen-bonded association in hydroxy compounds has been discussed in section 2.3; free O—H *str* bands are normally only seen in dilute solutions in nonassociating solvents. (Exceptions are highly hindered O—H groups or certain intramolecularly bonded *o*-hydroxybenzyl alcohols.)

2.10.1 ALCOHOLS (CHART 3(i))
The infrared spectrum of 1-butanol is typical, showing all of the expected absorptions shown on the correlation chart (see figure 2.6): the dotted trace on this spectrum was recorded on a dilute solution in CCl$_4$, and free O—H *str* appears near 3600 cm^{-1}.

Classification of an alcohol as primary, secondary or tertiary can frequently be successful using the bands for the coupled vibrations C—O *str* and O—H *def*.

2.10.2 CARBOHYDRATES (CHART 3(ii))
There is no difficulty normally in classifying a carbohydrate as such, but little further detail can be generally extracted from their spectra.

2.10.3 PHENOLS (CHART 3(iii))
In addition to the bands shown in the charts, phenols show the characteristic C=C *str* and C—H *def* vibrations for aromatic residues (chart 1(i)). The

strong O—H *str* band may swamp the weaker C—H *str* band just above 3000 cm^{-1}. It is difficult to distinguish a phenol from an aryl alcohol from infrared evidence.

2.10.4   ETHERS (CHART 3(iv))

Ethers show only one characteristic band for C—O *str* (in addition to C—H and C—C vibration bands arising from alkyl or aryl groups in the molecule): while this band is easily identified in *known* ethers, certain identification of an ether in an unknown molecule is difficult because many other strong bands can appear in the region 1050–1300 cm^{-1}.

In highly unsymmetrical ethers, especially alkyl aryl ethers, the two C—O bonds couple to give antisymmetric and symmetric C—O *str* absorptions: for dialkyl or diaryl members the symmetric C—O *str* is infrared inactive and only the antisymmetric C—O *str* is seen.

Note that keto ethers (such as *p*-methoxyacetophenone) may be confused with esters, since identification of both types rests on identifying both C=O *str* and C—O *str*. Epoxides are a special case of cyclic ether, and in addition to their C—O *str* band they have a reasonably characteristic C—H *str* absorption near 3050 cm^{-1}, and two bands around 800 and 900 cm^{-1}, respectively.

2.11   NITROGEN COMPOUNDS (CHART 4)

The correlation chart contains most of the information necessary to assign principal bands associated with nitrogen functions not already met in section 2.9 (and chart 2).

2.11.1   AMINES (CHART 4(i))

Hydrogen bonding in amines leads to modification of both the symmetric and antisymmetric N—H *str* bands, but shifts on dilution are less than for O—H *str* in alcohols and phenols. The values given on the charts refer to condensed phases (liquid films, KBr discs, etc.) and in dilute solution free N—H *str* can be seen near 3500 cm^{-1}. The infrared spectrum of *o*-toluidine (figure 2.16) shows all the characteristics of a primary aromatic amine spectrum: N—H *str* (two bands only may sometimes be seen), N—H *def* (at 1620 cm^{-1}) and C—N *str* at 1280 cm^{-1} (not identifiable in alkyl amines).

2.11.2   IMINES AND ALDEHYDE-AMMONIAS (CHART 4(ii))

These are both uncommon classes and the chart contains the only useful diagnostic bands for them.

2.11.3   NITRO COMPOUNDS (CHART 4(iii))

The symmetric and antisymmetric N=O *str* bands constitute a reliable method for detecting nitro groups, particularly in aryl nitro compounds (where few other detection methods are simple or reliable).

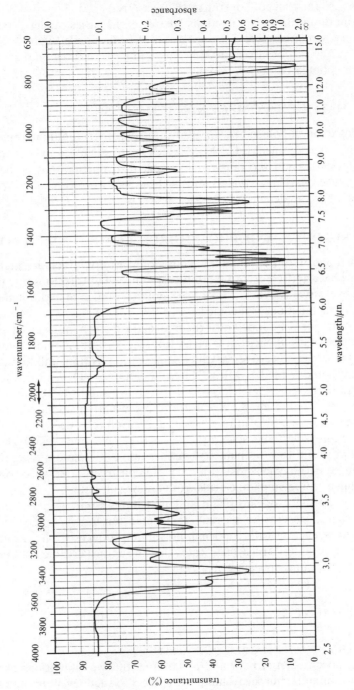

Figure 2.16 *Infrared spectrum of o-toluidine. Liquid film.*

2.11.4 NITRILES AND ISONITRILES (CHART 4(iv))

Without doubt these two functions are two of the easiest to detect: they each give one strong band, and the spectrum of benzonitrile (figure 2.17) shows clearly C≡N *str* for nitriles at 2250 cm^{-1}, in a spectral region that is usually sparsely populated.

2.12 HALOGEN COMPOUNDS (CHART 5)

The detection of halogen by infrared spectroscopy is totally unreliable. The most useful assignment is C—X *str*, which for most environments appears at frequencies beyond the range of low-cost instruments (that is, below 650 cm^{-1}). The chart nevertheless shows these band positions for the sake of interest and completeness.

2.13 SULPHUR AND PHOSPHORUS COMPOUNDS (CHART 6)

A wide range of sulphur-containing functional groups is included in the chart, mainly characterised according to the presence of absorptions due to S—H, S=O, SO$_2$, SO$_3$, C—S, C=S, or S—S bonds. A much more limited range of phosphorus functions is included.

Very little need be added over the information given in the chart, and only one representative spectrum is given (toluene-*p*-sulphonamide, figure 2.18). The S=O *str* bands are remarkably constant in position, and for sulphones and sulphonic acid derivatives, containing O=S=O groups, the strong antisymmetric and symmetric S=O *str* bands are highly reliable absorptions. Compounds containing the S—H group are usually recognised as such by their smells, before their infrared spectra are even recorded.

All amino sulphonic acids exist as zwitterions, since —SO$_3$H is sufficiently powerful to protonate aryl amines and alkyl amines alike (contrast aryl amino carboxylic acids). It is very difficult by infrared spectroscopy to distinguish amino sulphonic acids from amine sulphates.

The importance of phosphate and phosphite esters as plasticisers, fuel additives and insecticides warrants their inclusion in the correlation charts. The Wittig reaction with triphenylphosphine produces triphenylphosphine oxide as byproduct: it is useful to know that the unmistakeable P=O *str* band for this compound appears at 1190 cm^{-1}.

SUPPLEMENT 2

2S.1 QUANTITATIVE INFRARED ANALYSIS

Infrared spectra can provide a means of quantitative estimations of components in a mixture, but the technique is subject to such severe limitations that other analytical techniques (gas chromatography, ultraviolet spectroscopy)

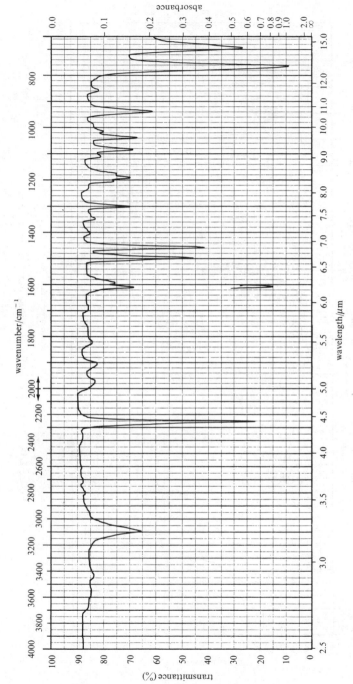

Figure 2.17 *Infrared spectrum of benzonitrile. Liquid film.*

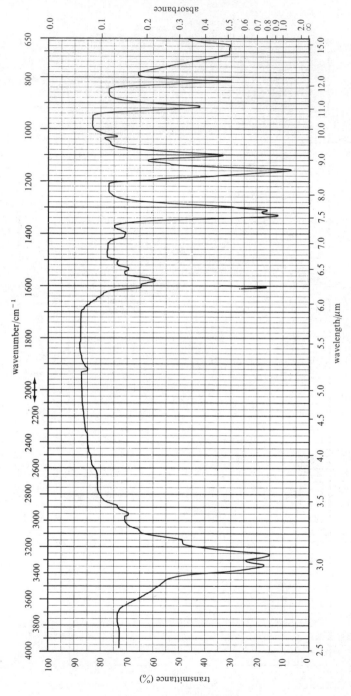

Figure 2.18 *Infrared spectrum of toluene-p-sulphonamide. KBr disc.*

should always be alternatively evaluated. As a successful example, the method can be used to measure small amounts of the isomeric *trans* acid in oleic acid (*cis*-$CH_3(CH_2)_7CH{=}CH(CH_2)_7CO_2H$), since the *trans* isomer has an absorption at 983 cm^{-1}, which is absent in the *cis* form.

2S.1.1 Absorbance

Quantitative analyses are best carried out using standard solutions, and for solutions the Beer–Lambert law should hold. The *absorbance* A of a solution is given by $A = \varepsilon cl$, where c is the molar concentration of the solute and l is the path length (in cm): ε is the *molar absorptivity*, and is constant for the solute causing the absorbance.

As we have seen in section 2.4, most infrared spectrometers measure band intensities linearly in per cent transmittance, $\%T$, and therefore logarithmically in absorbance A. Not only does this make it inconvenient to relate band intensity to concentration (which is related linearly with A but inversely and logarithmically with $\%T$), but slight inaccuracies in the measurement of $\%T$ may correspond to very large inaccuracies in the measurement of A. This is particularly true when $\%T$ approaches 100, and acceptable accuracy in the measurement of A can only be achieved when the band intensity lies between 30 and 60 $\%T$.

Absorbance is defined as $\log(I_0/I)$: in figure 2.19 we can theoretically measure I_0 for peak P as the distance AB, and thus I corresponds to BC. Since, however, there is always background absorption, and the spectrum never reaches $100\%T$, more accurate results are obtained using the *tangent baseline* DE. For peak Q, absorbance is given by

$$A = \log(I_0/I) = \log(FG/GH)$$

2S.1.2 Slit widths

The measured absorbance of a peak depends on the slit width at that wavelength—the wider the slit width the broader and flatter will be the band appearance, because of the intrusion through the slit of neighbouring wavelengths at which no absorption is taking place. In practice, reproducibility in the measurement of absorbance can only be achieved if the slit width is less than one fifth of the band width at half height.

2S.1.3 Path lengths

Solution-cell path lengths cannot accurately be inferred from the spacer thickness, and are best measured by the measurement of *interference fringes*. In figure 2.20, the beam 1 is transmitted directly, while beam 2 suffers double internal reflection at the internal surfaces of the cell; the consequent delay in beam 2 leads to reinforcement and cancellation of the resultant beam at appropriate wavelength values, and a scan over the wavelength range of the instrument (through the empty cell) leads to an interference pattern as shown.

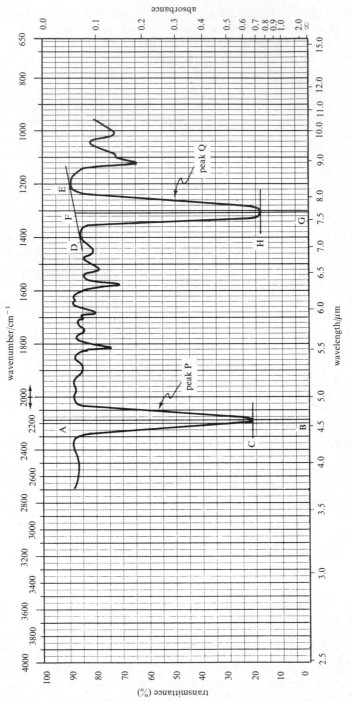

Figure 2.19 *Measurement of absorbance in quantitative infrared analysis.*

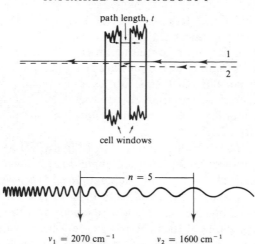

$$v_1 = 2070 \text{ cm}^{-1} \qquad v_2 = 1600 \text{ cm}^{-1}$$
$$t = 0.5n/(v_1 - v_2)$$
$$= 0.5 \times 5/(2070 - 1600) \text{ cm}^{-1}$$
$$= 5.3 \times 10^{-3} \text{ cm}$$

Figure 2.20 *Interference fringes in an empty solution cell. Measurement of path length.*

The delay in beam 2 is equal to $2t$ (t = cell thickness), and when $2t$ equals an integral number of wavelengths, reinforcement leads to a fringe maximum; at $2t - \frac{1}{2}$ integral number of wavelengths, cancellation leads to a fringe minimum. By choosing two reinforcement maxima (v_1 and v_2) and counting the number of intervening interference fringes between them (n), the cell thickness in this example is given by $t = 0.5n/(v_1 - v_2)$.

Interference fringes can also be seen in the spectra of polystyrene film (figures 2.9 and 2.10), between 2000 and 2800 cm^{-1}: film thickness can be calculated as above using the modified equation $t = 0.5n/(v_1 - v_2)\mu$, where μ is the refractive index of the film material. (For polystyrene $\mu = 1.6$.)

2S.1.4 Molar absorptivity

Molar absorptivity ε can only be measured accurately when all the above parameters are taken into account; coupled with the instrumental difficulty of producing infrared sources and detectors which are linear over the wavelength range, it is clear that accurate ε values are rarely achieved.

The most successful quantitative infrared work is done on a routine repetitive analytical basis using the same machine, the same machine settings of slit widths and scan speeds etc., and the same sample and reference cells. Even more successful is the preparation of standard solutions of pure authentic specimens, from which a calibration curve of $\%T$ against solute concentration can be drawn up for a given set of machine parameters. The concentration of solute in an unknown sample can then be most accurately measured.

2S.2 ATTENUATED TOTAL REFLECTANCE (A.T.R.) AND MULTIPLE INTERNAL REFLECTANCE (M.I.R.)

It is frequently necessary to measure the infrared spectrum of a material for which the normal sampling techniques are inapplicable; examples of such material are polymer films or foams, fabrics, thick pastes, coatings such as paint films or paper glazes, printing inks on metal, etc. The spectrum can be recorded using a modified reflectance technique which depends on the total internal reflectance of light.

Provided the refractive index of the prism 1 (figure 2.21(a)) is greater than that of the sample medium 2, and if the angle θ is greater than the critical angle, the infrared light beam 2 will suffer *total internal reflectance* at the interface between 1 and 2. The light beam must in fact travel a very short distance (a few μm) along medium 2 before re-emerging into 1, and if medium 2 absorbs some of the light then the re-emerging beam will be attenuated (weakened), rather than *totally* reflected. The principle can be easily extended to produce the infrared absorption spectrum of 2, the technique being called *attenuated total reflectance*, A.T.R.

Because of the very short path length within medium 2 (the sample) on a single reflection, a very weak spectrum results unless multiple reflectance can be arranged as in figure 2.21(b). This whole assemblage can be inserted into the sample beam of the infrared spectrometer, achieving up to 25 reflections and producing spectra comparable to normal transmission

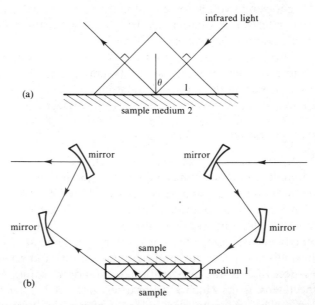

Figure 2.21 *Attachments for (a) attenuated total reflectance (A.T.R.) and (b) multiple internal reflectance (M.I.R.) measurements.*

spectra. This *multiple internal reflectance* (M.I.R.) is preferred to simple A.T.R. The alternative name *frustrated multiple internal reflectance* (F.M.I.R.) is also applied.

The success of the method depends on using a medium 1 which is transparent to infrared light, and has a high refractive index (μ, 2.5–3.5). Crystals normally used are germanium, silver chloride, or KRS-5 (mixed ThBr and ThI ($\mu = 2.4$) transmitting to 40 μm). The intensity of the spectrum depends on the efficiency of contact between the crystal and the sample, on the contact area and on the angle of reflectance. In double-beam spectrometers, because the A.T.R. unit increases the path length in the sample beam, a corresponding increase in the reference beam has to be provided by a second (blank) A.T.R. unit. Less successfully, the reference-beam intensity can be reduced by a comb attenuator, but this does not cope with interference from atmospheric absorptions, due to atmospheric CO_2 and H_2O.

2S.3   LASER-RAMAN SPECTROSCOPY

2S.3.1   The Raman effect

When a compound is irradiated with an intense exciting light source (mercury arc or laser beam) some of the light is scattered as shown in figure 2.22.

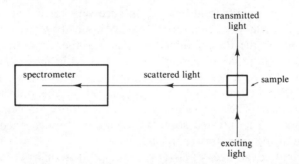

Figure 2.22   *Origin of Raman spectrum in scattered light.*

If the scattered light is passed into a spectrometer, we can obviously detect in it the frequency of the exciting radiation as a strong line (the so-called *Rayleigh line*) but the Indian physicist C. V. Raman recorded in 1928 that additional frequencies are also present in the scattered beam; these frequencies are symmetrically arrayed above and below the frequency of the Rayleigh line, and the *differences* between the Rayleigh-line frequency and the frequencies of the weaker *Raman lines* correspond to the vibrational frequencies present in the molecules of the sample. By measuring these *frequency differences* we can obtain information about the vibrational frequencies within the sample molecules. For example, we may observe Raman lines at $\pm$ 750 cm^{-1}, on either side of the Rayleigh line; the molecule therefore possesses a vibrational mode of this frequency. The plot of all the Raman frequency shifts against their intensities is a Raman spectrum, which is very similar to the vibrational spectrum produced in infrared spectroscopy.

Raman spectroscopy mainly utilised mercury arcs as the source of the exciting radiation until the development of the laser, and the era of routine laser-Raman work has only now arrived. Not only is laser light more intense and more cohesive, but it enables longer wavelengths to be used for excitation, eliminating many problems of sample decomposition.

2S.3.2   Comparison of infrared and Raman spectra

An infrared absorption arises when infrared light (electromagnetic) interacts with a fluctuating *dipole* within the molecule. The Raman effect is not an absorption effect like infrared, but depends on the *polarisability* of a vibrating group, and on its ability to interact and couple with an exciting radiation whose frequency does not match that of the vibration itself. A given vibrational mode within a molecule may lead to fluctuating changes in the dipole (an *infrared active* vibration), but may not necessarily lead to changes in the polarisability (a *Raman active* vibration) the converse also being true. Many absorptions that are weak in infrared (for example, the stretching vibrations of symmetrically substituted C=C and C≡C) give strong absorptions in Raman spectra.

The 'change in polarisability' which is necessary for a Raman signal to be seen is, to a close approximation, the 'ease of imposition of a dipole', but the dipole involved in the Raman signal may not be the same dipole involved in an associated infrared absorption signal. Indeed in molecules with high symmetry (for example benzene) the infrared-active vibrations that lead to dipole changes and those that lead to changes in the polarisability are mutually exclusive. The two techniques are therefore complementary, and figure 2.23 shows the infrared spectrum of indene (upper trace) compared with its Raman spectrum (lower trace).

The principal advantages of laser-Raman spectroscopy over infrared are its increased sensitivity (sample sizes as small as several nanograms), the ease of sample preparation (powders or solutions may be used), and the fact that water makes an ideal solvent (since its Raman spectrum consists essentially of only one broad weak band at 3654 cm^{-1}). The application to aqueous biological samples is extensive, and includes drugs analyses in body fluids and studies of aqueous solutions of amino acids. Organometallic complexes can also be easily studied in aqueous solution.

2S.4   MICRO METHODS

2S.4.1   General methods

In principle there are few difficulties in recording satisfactory infrared spectra from microgram samples, although technique becomes more exacting as in all

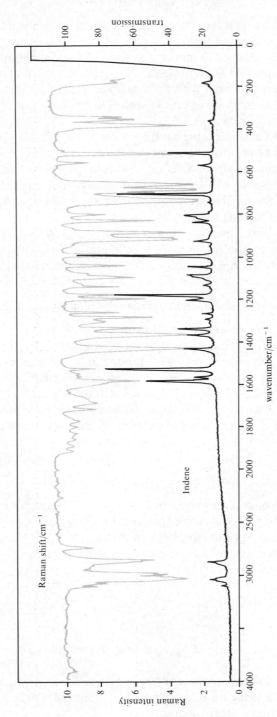

Figure 2.23 *Infrared and laser-Raman spectra of indene compared (infrared spectrum above).*

microanalytical work. It is far easier to handle small amounts of solids than liquids, and KBr microdiscs form the mainstay of microsampling methods. Conventional KBr discs are 13 mm in diameter, and microdies are available for pressing 1.5 mm and 0.5 mm discs. Alternatively, cardboard discs of 13 mm diameter are produced with 1.5 mm holes in their centres; this small hole is filled with the mixture of sample and KBr, and the whole pressed in the normal 13-mm KBr die.

With such small apertures, the sample beam is substantially attenuated and the reference beam must be correspondingly attenuated (for example, using a comb attenuator) to restore double-beam efficiency. The ideal is to use a beam condenser (a KBr lens) in the sample compartment to focus the entire beam onto the small aperture of the microdisc: this obviates the need for reference-beam attenuation.

Where microsamples have to be recovered from solution, this can be achieved by grinding the solution with KBr and allowing the solvent to evaporate. Disposable AgCl windows can also be used: the solution is dropped onto these thin sheets of AgCl and the solvent allowed to evaporate.

2S.4.2   Thin-layer chromatography/infrared spectrometry (T.L.C./i.r.)

Microsamples may be solvent extracted from the adsorbant on T.L.C. plates, and the resulting solution treated as above. Contamination by the adsorbant (commonly silica gel) is difficult to avoid, but can be circumvented using a patented porous triangle of KBr. The T.L.C. spot is scraped off and added to a suitable solvent contained in a tall phial closed at the top with a lid having a hole in its centre. The triangle of KBr is placed tip uppermost in the solvent and in the manner of T.L.C. the solvent travels up the triangle and evaporates at the tip through the hole in the lid. In so doing the solute is eluted up to the tip of the KBr triangle, which can be cut off and pressed into a microdisc.

2S.4.3   Gas chromatography/infrared spectrometry (G.C./i.r.)

All the analytical potential of gas chromatography can be coupled to infrared analysis by the direct condensation of peaks from a preparative gas chromatography column into AgCl microcells. Rapid loss of volatile samples can to some extent be overcome by using a rapid-scanning spectrometer.

FURTHER READING

MAIN TEXTS

L. J. Bellamy, *The Infra-red Spectra of Complex Molecules*, Methuen, London (1958).

L. J. Bellamy, *Advances in Infrared Group Frequencies*, Methuen, London (1968). (These two texts are undoubtedly the standard works for organic applications of infrared spectroscopy.)

A. D. Cross, *Introduction to Practical Infrared Spectroscopy*, Butterworths, London (2nd edn, 1964).

J. H. van der Maas, *Basic Infrared Spectroscopy*, Heyden, London (2nd edn, 1972).

SPECTRA CATALOGUES

N. Luff, *D.M.S. Working Atlas in Infrared*, Butterworths, London (1970). (Very valuable collection grouped according to functional class, etc.)

R. Mecke and J. Langenbucher, *Infrared Spectra of Selected Chemical Compounds*, Heyden, London (1965). (A catalogue of around 1800 spectra indexed by name and molecular formula.)

Documentation of Molecular Spectroscopy (*D.M.S.*), Butterworths, London. (Almost too comprehensive, and more research-oriented.)

Charles J. Pouchert, *Aldrich Library of Infrared Spectra*, Aldrich Chemical Co. Inc., Milwaukee (1971).

SUPPLEMENTARY TEXTS

S. K. Freeman, *Applications of Laser Raman Spectroscopy*, Wiley, London (1973).

M. C. Tobin, *Laser Raman Spectroscopy*, Wiley-Interscience, Chichester (1971).

3

NUCLEAR MAGNETIC RESONANCE SPECTROSCOPY†

As is implied in the name, nuclear magnetic resonance (or n.m.r.) is concerned with the magnetic properties of certain atomic nuclei, notably the nucleus of the hydrogen atom—the proton. The growth of n.m.r. since the early sixties has been nothing short of explosive and has led to highly sophisticated instrumental techniques, which in turn have led to complex organic structural investigations and the elucidation of mechanistic and stereochemical detail hitherto inaccessible.

Studying an organic molecule by n.m.r. spectroscopy enables us to record differences in the magnetic properties of the various nuclei present, and to deduce in large measure what are the positions of these nuclei within the molecule. We can, for proton n.m.r., deduce how many different kinds of hydrogen environments there are in the molecule, and also which hydrogen atoms are present on neighbouring carbon atoms. We can also measure how many hydrogen atoms are present in each of these environments.

The n.m.r. spectrum of toluene, $C_6H_5CH_3$, represents an extremely

† The following n.m.r. spectra are reproduced from *High Resolution NMR Spectra Catalog*, with permission of the publishers, Varian Associates, Palo Alto, California, U.S.A.: figures 3.4, 3.11, 3.12, 3.17, 3.18, 3.19, 3.20, 3.24, and the ¹H n.m.r. spectrum of menthol in figure 3.37.

Figures 3.31 and 3.35 are reproduced from *NMR Quarterly*, with permission of the publishers, Perkin–Elmer Limited, Beaconsfield, Buckinghamshire. The ¹³C n.m.r. spectrum of menthol in figure 3.37 is reproduced with permission of Jeol U.K. Limited, Colindale, London.

All ¹n.m.r. spectra were recorded at 60 MHz unless stated otherwise.

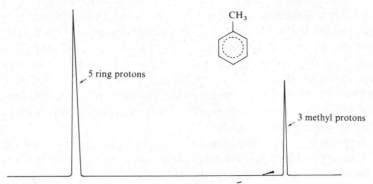

Figure 3.1 *Diagrammatic n.m.r. spectrum of toluene,* $C_6H_5CH_3$, *showing two signals in the intensity ratio 5:3.*

simple example of an n.m.r. spectrum, and is shown in diagrammatic form in figure 3.1.

Toluene has two groups of hydrogen atoms—the methyl hydrogens and the ring hydrogens. *Two signals* appear on the spectrum, corresponding to these *two different chemical and magnetic environments.*

The areas under each signal are in the ratio of the number of protons in each part of the molecule, and measurement will show that *the ratio of these areas is 5:3.*

3.1 THE N.M.R. PHENOMENON

3.1.1 THE SPINNING NUCLEUS
The nucleus of the hydrogen atom (the proton) behaves as a tiny spinning bar magnet, and it does so because it possesses both electric charge and mechanical spin; any spinning charged body will generate a magnetic field, and the nucleus of hydrogen is no exception.

3.1.2 THE EFFECT OF AN EXTERNAL MAGNETIC FIELD
Like all bar magnets, the proton will respond to the influence of an external magnetic field, and will tend to align itself with that field, in the manner of a compass needle in the earth's magnetic field. Because of quantum restrictions which apply to nuclei but not to compass needles (see section 3.2), the proton can only adopt two orientations with respect to an external magnetic field—*either aligned with the field* (the lower energy state) or *opposed to the field* (the higher energy state). We can also describe these orientations as *parallel* with or *antiparallel* with the applied field.

3.1.3 PRECESSIONAL MOTION
Because the proton is behaving as a *spinning* magnet, not only can it align itself with or oppose an external magnetic field, but it will move in a characteristic way under the influence of the external magnet.

Consider the behaviour of a spinning top: as well as describing its *spinning* motion, the top will (unless absolutely vertical) also perform a slower waltz-like motion, in which the spinning axis of the top moves slowly around the vertical. This is *precessional* motion, and the top is said to be *precessing* around the vertical axis of the earth's gravitational field. The precession arises from the interaction of spin, that is gyroscopic motion, with the earth's gravity acting vertically downwards. Only a *spinning* top will precess; a static top will merely fall over.

As the proton is a *spinning* magnet, it will, like the top, precess around the axis of an applied external magnetic field, and can do so in two principal orientations, either aligned with the field (low energy) or opposed to the field (high energy). This is represented in figure 3.2, where B_0 is the external magnetic field.

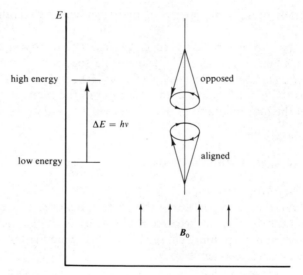

Figure 3.2   *Representation of precessing nuclei, and the ΔE transition between the aligned and opposed conditions.*

3.1.4   PRECESSIONAL FREQUENCY

The spinning frequency of the nucleus does not change, but the speed of precession does. The *precessional frequency*, ν, is directly proportional to the strength of the external field B_0: that is

$$\nu \propto B_0$$

This is one of the most important relationships in n.m.r. spectroscopy, and it is restated more quantitatively in section 3.2.

As an example, a proton exposed to an external magnetic force of 1.4 T ($\equiv$ 14 000 gauss) will precess $\approx$ 60 million times per second, so that $\nu = 60$

MHz. For an external field to 2.3 T, v is $\approx$ 100 MHz, and at 5.1 T v is $\approx$ 220 MHz. (Strictly, the tesla is a measure of magnetic flux density, not field strength.)

3.1.5 ENERGY TRANSITIONS

We have seen that a proton, in an external magnetic field of 1.4 T, will be precessing at a frequency of $\approx$ 60 MHz, and be capable of taking up one of two orientations with respect to the axis of the external field—aligned or opposed, parallel or antiparallel.

If a proton is precessing in the *aligned* orientation, it can absorb energy and pass into the *opposed* orientation; subsequently it can lose this extra energy and relax back into the aligned position. If we irradiate the precessing nuclei with a beam of radiofrequency energy of the correct frequency, the low-energy nuclei may absorb this energy and move to a higher energy state. The precessing proton will only absorb energy from the radiofrequency source if the precessing frequency is the same as the frequency of the radiofrequency beam; when this occurs, the nucleus and the radiofrequency beam are said to be *in resonance*; hence the term *nuclear magnetic resonance*.

The simplest n.m.r. experiment consists in exposing the protons in an organic molecule to a powerful external magnetic field; the protons will precess, although they may not all precess at the same frequency. We irradiate these precessing protons with radiofrequency energy of the appropriate frequencies, and promote protons from the low-energy (aligned) state to the high-energy (opposed) state. We record this absorption of energy in the form of an n.m.r. spectrum, such as that for toluene in figure 3.1.

3.2 THEORY OF NUCLEAR MAGNETIC RESONANCE

The only nuclei that exhibit the n.m.r. phenomenon are those for which the spin quantum number I is greater than 0: the spin quantum number I is associated with the mass number and atomic number of the nuclei as follows

Mass number	Atomic number	Spin quantum number
odd	odd or even	$\frac{1}{2}, \frac{3}{2}, \frac{5}{2} \cdots$
even	even	0
even	odd	$1, 2, 3, \ldots$

The nucleus of ^{1}H, the proton, has $I = \frac{1}{2}$, whereas ^{12}C and ^{16}O have $I = 0$ and are therefore nonmagnetic. If ^{12}C and ^{16}O had been magnetic, the n.m.r. spectra of organic molecules would have been much more complex.

Other important magnetic nuclei that have been studied extensively by n.m.r. are ^{11}B, ^{13}C, ^{19}F and ^{31}P. (^{13}C, ^{19}F and ^{31}P n.m.r. are discussed in supplement 3.) Both deuterium (^{2}H) and nitrogen (^{14}N) have $I = 1$, and the consequences of this observation will become apparent later (see Hetero-nuclear coupling, section 3.9.4.).

Under the influence of an external magnetic field, a magnetic nucleus can take up different orientations with respect to that field; the number of possible orientations is given by $(2I + 1)$, so that for nuclei with spin $\frac{1}{2}$ (^{1}H, ^{13}C, ^{19}F, etc.) only two orientations are allowed. Deuterium and ^{14}N have $I = 1$ and so can take up three orientations: these nuclei do not simply possess magnetic *dipoles*, but rather possess electric *quadrupoles*. Nuclei possessing electric quadrupoles can interact both with magnetic and electric field gradients, the relative importance of the two effects being related to their magnetic moments and electric quadrupole moments, respectively.

In an applied magnetic field, magnetic nuclei like the proton precess at a frequency v, which is proportional to the strength of the applied field. The exact frequency is given by

$$v = \frac{\mu \, \beta_N \, B_0}{hI}$$

where B_0 = strength of the applied external field experienced by the proton,
$\quad I$ = spin quantum number,
$\quad h$ = Planck's constant (6.626×10^{-34} J s),
$\quad \mu$ = magnetic moment of the particular nucleus,
$\quad \beta_N$ = nuclear magneton constant.

Typical approximate values for v are shown in table 3.1 for selected values of field strength B_0, for common magnetic nuclei.

The strength of the signal, and hence the sensitivity of the n.m.r. experiment for a particular nucleus is related to the magnitude of the magnetic moment μ. The magnetic moments of ^{1}H and ^{19}F are relatively large, and detection of n.m.r. with these nuclei is fairly sensitive. The magnetic moment for ^{13}C is about one quarter that of ^{1}H, and that of ^{2}H is roughly one third the moment of ^{1}H; these nuclei are less sensitively detected in n.m.r. (In contrast, the magnetic moment of the free electron is nearly 700 times that of ^{1}H, and

Table 3.1 Precessional frequencies (in MHz) as a function of increasing field strength

B_0/tesla	1.4	2.1	2.3	5.1	5.8	7.1
Nucleus						
^{1}H	60	90	100	220	250	300
^{2}H	9.2	13.8	15.3	33.7	38.4	46.0
^{13}C	15.1	22.6	25.2	55.0	62.9	75.5
^{14}N	4.3	6.5	7.2	15.8	17.9	21.5
^{19}F	56.5	84.7	93.0	206.5	203.4	282.0
^{31}P	24.3	36.4	40.5	89.2	101.5	121.5
(Free electron)	3.9×10^4					

resonance phenomena for free radicals can be studied in extremely dilute solutions; see section 3S.5 for a discussion on electron spin resonance.)

Even with very large external magnetic fields, up to 7.1 T, the energy difference $\Delta E = hv$ is very small, up to 300 MHz. Because the difference is so small (of the order 10^{-4} kJ mol^{-1}) the populations of protons in the two energy states are nearly equal; the calculated Boltzmann distribution for ^{1}H at 1.4 T and at room temperature shows that the lower energy state has a nuclear population only about 0.001 per cent greater than that of the higher energy state. The relative populations of the two spin states will change if energy is supplied of the correct frequency to induce transitions upwards or downwards.

What happens when protons absorb 60 MHz radiofrequency energy?

Nuclei in the lower energy state undergo transitions to the higher energy state; the populations of the two states may approach equality, and if this arises no further net absorption of energy can occur, and the observed resonance signal will fade out. We describe this situation in practice as *saturation* of the signal. In the recording of a normal n.m.r. spectrum, however, the populations in the two spin states do not become equal, because higher-energy nuclei are constantly returning to the lower energy spin state.

How can the nuclei lose energy and undergo transitions from the high- to the low-energy state?

The energy difference, ΔE, can be re-emitted as 60 MHz energy, and this can be monitored by a radiofrequency detector as evidence of the resonance condition having been reached. Of great importance, however, are two *radiationless* processes, which enable high-energy nuclei to lose energy.

The high-energy nucleus can undergo energy loss (or *relaxation*) by transferring ΔE to some electromagnetic vector present in the surrounding environment. For example, a nearby solvent molecule, undergoing continuous vibrational and rotational changes, will have associated electrical and magnetic changes, which might be properly oriented and of the correct dimension to absorb ΔE. Since the nucleus may be surrounded by a whole array of neighbouring atoms, either in the same molecule or in solvent molecules, etc., this relaxation process is termed *spin–lattice relaxation*, where *lattice* implies the entire framework or aggregate of neighbours.

A second relaxation process involves transferring ΔE to a neighbouring nucleus, provided that the particular value of ΔE is common to both nuclei: this mutual exchange of spin energy is termed *spin–spin relaxation*. While one nucleus loses energy, the other nucleus gains energy, so that no net change in the populations of the two spin states is involved.

The rates of relaxation by these processes are important, and in particular the rate of spin–lattice relaxation determines the rate at which *net* absorption of 60 MHz energy can occur.

The mean half-life of the spin–lattice relaxation process is designated T_1, and that of the spin–spin relaxation process T_2. If T_1 and T_2 are small, then the lifetime of an excited nucleus is short, and it has been found that this gives rise to very broad absorption lines in the n.m.r. spectrum. If T_1 and T_2 are large, perhaps of the order 1 second, then sharp spectral lines arise.

For nonviscous liquids (and that includes solutions of solids in nonviscous solvents) molecular orientations are random, and transfer of energy by spin–lattice relaxation is inefficient. In consequence T_1 is large, and this is one reason why sharp signals are obtained in n.m.r. studies on nonviscous systems.

The important relationship between relaxation times and line broadening can be understood qualitatively by using the uncertainty principle in the form $\Delta E \cdot \Delta t \approx h/2\pi$, or, since $E = hv$, $\Delta v \cdot \Delta t \approx 1/2\pi$. Expressed verbally, the product $\Delta v \cdot \Delta t$ is constant, and if Δt is large (that is the lifetime of a particular energy state is long) then Δv must be small (that is the uncertainty in the measured frequency must be small, so that there is very little 'spread' in the frequency, and line-widths are narrow). Conversely if Δt is small (fast relaxation) then Δv must be large, and broad lines appear.

The ^{14}N nucleus possesses an electric quadrupole (see page 80), and is therefore able to interact with both electric and magnetic field gradients, which cause the nucleus to tumble rapidly: spin–lattice relaxation is highly effective, and therefore T_1 is small. Since T_1 is small, n.m.r. signals for the ^{14}N nucleus are very broad indeed, and for the same reason the n.m.r. signals for most protons attached to ^{14}N (in N—H groups) are broadened. (See also under Heteronuclear coupling, page 113).

3.3   CHEMICAL SHIFT AND ITS MEASUREMENT

For a field strength of 1.4 T, we have seen that protons have a precessional frequency of ≈ 60 MHz. The precessional frequency of all protons in the same external applied field is not, however, the same, and the precise value for any one proton depends on a number of factors. Historically, this was first observed by Packard in 1951; he was able to detect three different values for the precessional frequencies of the protons in ethanol, and the realisation that these corresponded to the three different chemical environments for the protons in ethanol (CH_3, CH_2 and OH) marked the beginning of n.m.r. as a tool of the organic chemist. Because the shift in frequency depended on chemical environment, this gave rise to the term *chemical shift*. An even simpler example is the n.m.r. spectrum of toluene shown in figure 3.1. The protons in toluene are in two different chemical environments, and two signals appear on the spectrum, corresponding to two different precessional frequencies.

We say that the two groups of protons have different *chemical shift positions* on the spectrum.

3.3.1 Measurement of Chemical Shift—Internal Standards

To measure the precessional frequency of a group of nuclei in absolute frequency units is extremely difficult and rarely required. More commonly the *differences* in frequency are measured with respect to some reference group of nuclei. For protons, the universally accepted reference is tetramethylsilane, TMS

$$H—\underset{\underset{H}{|}}{\overset{\overset{H}{|}}{Si}}—H \qquad CH_3—\underset{\underset{CH_3}{|}}{\overset{\overset{CH_3}{|}}{Si}}—CH_3 \qquad CH_3—\underset{\underset{CH_3}{|}}{\overset{\overset{CH_3}{|}}{Si}}—CH_2CH_2CH_2SO_3{}^-Na^+$$

silane   tetramethylsilane   sodium salt of
                      (TMS)   3-(trimethylsilyl)-propanesulphonic acid

TMS is chosen because it gives an intense sharp signal even at low concentrations (having 12 protons in magnetically equivalent positions); the signal arises on the n.m.r. spectrum well clear of most common organic protons (for reasons we shall meet in section 3.4); it is chemically inert and has a low boiling point, so that it is easily removed from a recoverable sample of a valuable organic compound; it is soluble in most organic solvents, and can be added to the sample solution (≈ 1 per cent) as an *internal standard*.

TMS is not soluble in water or in D_2O; for solutions in these solvents the sodium salt of 3-(trimethylsilyl)-propanesulphonic acid is used.

The choice of standards against which to measure the precessional frequency of other nuclei such as ^{13}C and ^{19}F is not so well decided, although for ^{13}C TMS is now generally accepted, and $CFCl_3$ is most widely used for ^{19}F. The n.m.r. spectroscopy of nuclei other than ^1H is discussed more fully in supplement 3.

3.3.2 Measurement of Chemical Shift—The n.m.r. Spectrometer

The basic features of the instrumentation needed to record an n.m.r. spectrum are a magnet, a radiofrequency source and a detection system to indicate that energy is being transferred from the radiofrequency beam to the nucleus. Such an arrangement is shown schematically in figure 3.3.

Magnet strengths and frequencies depend on the various factors discussed above, but typical practical parameters for proton n.m.r. will be used for this discussion.

The sample is placed in a glass tube (for example 5 mm diameter) between the pole faces of a magnet of field strength 1.4 T. A radiofrequency source feeds energy at 60 MHz into a coil wound around the sample tube, and the radiofrequency detector is tuned to 60 MHz. If the nuclei in the sample do not resonate with the 60 MHz source, the detector will only record a weak signal coming directly from the source coil to the detector coil. An increased signal will be detected if nuclei in the sample resonate with the source,

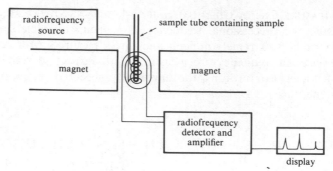

Figure 3.3   *Basic features of an n.m.r. spectrometer.*

since energy will be transferred from the source, via the nuclei, to the detector coil. The output from the detector can be fed to a cathode-ray oscilloscope or to a strip chart recorder after amplification, etc.

Because of chemical shift, not all protons come to resonance at exactly 60 MHz in a field of 1.4 T, and we must have provision in the instrument for detecting these different frequencies. An easy method in theory is to hold the magnetic field steady, and use a tunable radiofrequency source (or 'synthesiser') which can scan over the nearby frequencies until the various nuclei come to resonance in turn as their precessional frequencies are matched by the scanning source. Such an instrument, operating on *frequency sweep*, is difficult to construct with sufficient frequency accuracy, although there are commercial models marketed using this technique.

It is in practice cheaper and easier to make use of the fact that $v \propto B_0$. By holding the radiofrequency source steady at 60 MHz and varying B_0, we can increase or decrease the precessing frequencies of all the nuclei, until each, in turn, reaches 60 MHz and comes to resonance with the radiofrequency source. In such an instrument, the field strength is varied by fixing small electromagnets to the pole faces of the main magnet; by increasing the current flowing through these electromagnets the total field strength is increased. As the field strength increases, so the precessional frequency of each proton increases until resonance with the 60 MHz source takes place. This design of instrument is operating on *field sweep*, and the majority of commercial n.m.r. machines use this mode. The variable electromagnet coils are called *sweep coils*.

As each proton comes to resonance, the signal from the detector produces a peak on the chart; the n.m.r. spectrum is therefore a series of peaks plotted on an abscissa that corresponds to variations in field strength or (since $v \propto B_0$) frequency.

Differences in precessional frequencies are very small indeed, and modern high-resolution instruments can detect differences of the order 0.01 parts per million (p.p.m.). To obtain these accuracies in resolution, the radiofrequency source and the field strength of the magnet must be accurate to one part in 10^8 or better, and this remarkable accuracy is achieved by a number of means. Much of the accuracy is obtained by having additional contoured electromagnet coils on the pole faces of the main magnet (in addition to the sweep coils); these coils (*Golay coils*) can be 'tuned' to create specifically contoured magnetic fields, which compensate for any inhomogeneity in the main magnet's field. As a further refinement, any remaining magnet inhomogeneity in the horizontal plane can be averaged out if the sample tube is spun about its axis, so that the molecules in different parts of the sample tube experience the same *average* magnetic environment. The sample tube is mounted on a light turbine, and a jet of air is adjusted to provide a steady spinning rate of around 30 Hz.

Adjustment of magnet homogeneity may require weekly or even daily attention to the Golay coils. A standard test of homogeneity is the sharpness and symmetry of the so-called *ringing* pattern associated with the n.m.r. signal and seen under amplification in most sharply resolved proton n.m.r. signals. The ringing pattern is a beat phenomenon caused by the interaction of two close frequencies and has the appearance of a steady exponential decay of pen oscillations. At resonance, the nucleus sets up weak rotating vectors whose frequency alters as the field is swept; interaction of these vectors with the normal relaxation vectors produces the wiggle-beat pattern called ringing. The effect of ringing is seen clearly in figure 3.22: on the high-field side of each peak, the pen has oscillated above and below the base line.

Occasionally a strong signal in the spectrum will be flanked by two smaller signals equidistant from the main signal. This may be caused by oscillations set up by irregularities in the spinning of the sample tube; these oscillations will act as modulators, and produce the visible side-bands above and below the main signal frequency (by addition and subtraction of the modulating frequency). These *spinning side-bands* are always symmetrically displayed, but they vary in position with the spinning rate of the sample tube; they are not to be confused with the satellite side-bands caused by ^{13}C nuclei, as discussed in section 3S.3. Spinning side-bands are minimised by the use of high-precision sample tubes, and by the avoidance of excessive spinning rates.

Not all instruments make use of the two sets of radiofrequency coils shown in figure 3.3 although these *crossed coils* (so called because they are arranged orthogonally) give high signal-to-noise ratios and are widely used. It is possible to perform the n.m.r. experiment using only one radiofrequency coil with a radiofrequency bridge arrangement (similar to the Wheatstone bridge) to monitor out-of-balance changes in signal intensities as radio-frequency energy is absorbed by the nuclei.

It is easier to achieve accurate homogeneity over a *small area* of the magnet pole faces, and small sample sizes are advantageous (≈ 0.3 cm^3 of 10 per cent solutions are commonly used in proton work). This consideration is offset for nuclei with small magnetic moments, etc., where large sample sizes are demanded for detection sensitivity.

To record the n.m.r. spectra of nuclei other than ^{1}H, using the instrument above (field strength 1.4 T), requires a different radiofrequency source appropriate to the nucleus being examined (see table 3.1). For example, with a 1.4-T magnet, ^{19}F spectra can be recorded using a radiofrequency source at 56.5 MHz, and ^{13}C spectra require a 15.1 MHz source.

Instruments with a 1.4 T magnet are usually called 60 MHz instruments, although strictly this is only the frequency being used when ^{1}H spectra are being recorded. Similarly, 100 MHz and 220 MHz instruments would be more suitably called 2.3 T and 5.1 T instruments, respectively.

Field strengths of 1.4 T and 2.1 T (60 MHz and 90 MHz instruments) can be obtained either from permanent magnets or electromagnets. Field strengths of 2.3 T (100 MHz instruments) are now only obtained commercially from electromagnets. The really high field strengths of 5.1 T and 7.1 T (220 MHz and 300 MHz instruments) use superconducting magnets. When a metal conductor is cooled in liquid helium to a temperature of 4 K, electrical resistance vanishes and the metal becomes a 'superconductor'. An electromagnet immersed in liquid helium and using this principle can reach enormous field strengths, which were hitherto regarded as unattainable. Because of capital and running costs, very few 220 MHz instruments are available, and only one 300 MHz instrument is in commercial use as at January 1975.

3.3.3 MEASUREMENT OF CHEMICAL SHIFT—UNITS USED IN n.m.r.

The conventional n.m.r. spectrometer scans from low field to high field and the normal presentation of an n.m.r. spectrum has low-field values on the left and high on the right; the signal for TMS appears to the right of signals coming from the protons of most organic molecules. The n.m.r. spectrum of benzyl alcohol (PhCH$_2$OH) is shown in figure 3.4: from left to right, the signals correspond to the protons of the aromatic ring, the CH$_2$ group, the OH group and lastly the small TMS internal standard peak. We say that the protons of benzyl alcohol come to resonance *downfield of* TMS.

If we recorded the n.m.r. spectrum of benzyl alcohol using a frequency-sweep instrument it would have exactly the same appearance; the TMS protons would come to resonance at *lower* values of the sweeping frequency than the protons of benzyl alcohol. Increasing frequency is plotted from right to left.

The TMS protons are known to come to resonance at exactly 60 MHz when the field strength is 1.4092 T, and we can measure the positions of other signals in relation to TMS in parts per million (p.p.m.) downfield from TMS or p.p.m. higher frequency than TMS. Thus the difference between the TMS signal and the OH signal in benzyl alcohol is 2.4 p.p.m. (downfield from TMS or at higher frequency than TMS).

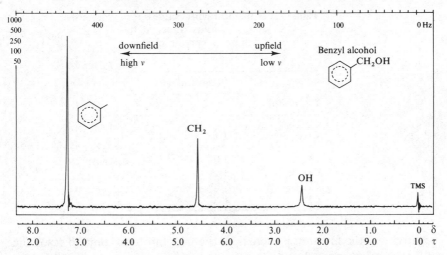

Figure 3.4   *N.M.R. spectrum of benzyl alcohol* $(PhCH_2OH)$. (CCl_4).

δ *units.* Chemical shift positions are frequently expressed in δ (delta) units, which are defined as differences, in p.p.m., from the TMS signal. The δ values for the benzyl alcohol protons are thus 2.4 δ for OH, 4.6 δ for CH_2 and 7.3 δ for the aromatic protons. The δ unit is a proportionality and therefore a dimensionless number. It is independent of field strength; therefore the above δ values for benzyl alcohol protons would be correct on a 100 MHz machine or on a 220 MHz machine, etc.

τ *units.* Using δ units, chemical shift values increase on the spectrum on going from right to left, while field strength and the terminology of upfield and downfield shifts implies increasing values from left to right. To overcome this trivial difficulty, Tiers suggested the τ (tau) unit

$$\tau = 10 - \delta$$

Thus the τ values for the protons in benzyl alcohol are 2.7 τ for the aromatic protons, 5.4 τ for the CH_2, and 7.6 τ for the OH proton. TMS appears at 10 τ by definition.

Although a slight majority of chemists use τ units, δ units are now internationally preferred and virtually all chemical shift data in this book are quoted in δ units.

3.4   FACTORS INFLUENCING CHEMICAL SHIFT

3.4.1   ELECTRONEGATIVITY—SHIELDING AND DESHIELDING

Table 3.2 shows the chemical-shift positions for CH_3 protons when a methyl group is attached to functions of increasing electronegativity. As the electro-

Table 3.2   Chemical shift values for CH_3 protons attached to groups of varying electronegativity

Compound	Chemical shift	
	δ	τ
$CH_3-\overset{\textstyle\vert}{\underset{\textstyle\vert}{Si}}-$	0.0	10.0
CH_3I	2.16	7.84
CH_3Br	2.65	7.35
CH_3Cl	3.10	6.90
CH_3F	4.26	5.74

negativity of the function is increased, the CH_3 protons come to resonance at higher δ (lower τ) values.

Why does it require a lower applied magnetic field to bring CH_3F to resonance than it does for CH_3Cl etc.? The explanation relates to the electron density around the ^1H nuclei.

Hydrogen nuclei are surrounded by electronic charge which to some extent *shields* the nucleus from the influence of the applied field B_0, and to bring a proton to resonance, the magnetic flux must overcome this *shielding effect*. In a magnetic field, the electrons around the proton are induced to circulate, and in doing so they generate a small secondary magnetic field, which acts in opposition (that is diamagnetically) to the applied field. The greater the electron density circulating around the proton, the greater the induced diamagnetic effect, and the greater the external field required to overcome the shielding effect. Electronegative groups, like fluorine in CH_3F, withdraw electron density from the methyl group (inductive effect) and this *deshielding effect* means that a lower value of the applied magnetic field is needed to bring the CH_3 protons to resonance. Fluorine is more electronegative than chlorine, so the protons in CH_3F appear at higher δ values than those in CH_3Cl.

Silicon is electropositive, and the opposite effect operates in, for example, TMS; silicon pushes electrons into the methyl groups of TMS by a $+I$ inductive effect, and this powerful *shielding effect* means that the TMS protons come to resonance at high field.

The effect of charged species on chemical shift values is very marked; protons adjacent to N^+ (as in quaternary ammonium ions R_4N^+) are very strongly deshielded (high δ values), while carbanionic centres act as powerful shielding influences (low δ values).

Much of the data in table 3.4 can be explained by electronegativity considerations, although we shall meet other powerful shielding and deshielding mechanisms in the following sections.

3.4.2 VAN DER WAALS DESHIELDING

In a rigid molecule it is possible for a proton to occupy a sterically hindered position, and in consequence the electron cloud of the hindering group will tend to repel, by electrostatic repulsion, the electron cloud surrounding the proton. The proton will be deshielded and appear at higher δ values than would be predicted in the absence of the effect. Although this influence is small (usually less than 1 p.p.m.) it must be borne in mind when predicting the chemical shift positions in overcrowded molecules such as highly substituted steroids or alkaloids.

3.4.3 ANISOTROPIC EFFECTS

The chemical shift positions (δ) for protons attached to C=C in alkenes is higher than can be accounted for by electronegativity effects alone. The same is true of aldehydic protons and aromatic protons, whereas alkyne protons appear at relatively low δ. Table 3.3 lists approximate δ values for these protons.

Table 3.3 Approximate chemical shift ranges for protons attached to anisotropic groups

Structure	Approximate chemical shift range/δ
—C(H)=O	9.5–10.0
C=C(H)	5–6
aromatic ring—H	7–8
—C≡C—H	1.5–3.5

The explanation is again collated with the manner in which electrons, in this case π-electrons, circulate under the influence of the applied field. The effect is complex, and can lead to downfield shifts (paramagnetic shifts) or upfield shifts (diamagnetic shifts). In addition, the effects are paramagnetic in certain directions around the π clouds, and diamagnetic in others, so that these effects are described as *anisotropic*, as opposed to *isotropic* (operating equally through space).

Alkenes. When an alkene group is so orientated that the plane of the double bond is at 90° to the direction of the applied field (as in figure 3.5) induced

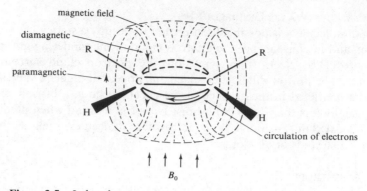

Figure 3.5　*Induced anisotropic magnetic field around an alkene group.*

circulation of the π-electrons generates a secondary magnetic field, which is diamagnetic around the carbon atoms, but paramagnetic (that is, it augments B_0) in the region of the alkene protons.

Where the direction of the induced magnetic field is parallel to the applied field B_0, the net field is greater than B_0. Protons in these zones require a lower value of B_0 to come to resonance, and therefore appear at lower field (higher δ values) than expected.

Any group held above or below the plane of the double bond will experience a *shielding effect*, since in these areas the induced field opposes B_0. In α-pinene one of the geminal methyl groups is held in just such a shielded position, and comes to resonance at significantly lower δ (higher field) than its twin. The third methyl group appears at higher δ (lower field) since it lies *in* the plane of the double bond and is thus *deshielded*.

α-pinene

In summary, we can divide the space around a double bond into two categories as shown in figure 3.6a. Deshielding occurs in the cone-shaped zones, and in these zones δ values will tend to be higher. Shielding is indicated by the + sign, and protons in these zones are shielded (lower δ values).

Carbonyl compounds. For the carbonyl group a similar situation arises, although the best representation of shielding and deshielding zones is slightly

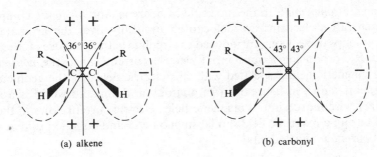

(a) alkene   (b) carbonyl

Figure 3.6   *Anisotropic shielding and deshielding around (a) alkene groups (b) carbonyl groups.*

different from the alkene pattern; see figure 3.6b. Two cone-shaped volumes, centred on the oxygen atom, lie parallel to the axis of the C=O bond; protons within these cones experience deshielding, so that aldehydic protons, and the formyl protons of formate esters, appear at high δ values. Protons held above or below these cones will come to resonance at lower δ values.

Alkynes. Whereas alkene and aldehydic protons appear at high δ values, alkyne protons appear around 1.5–3.5 δ. Electron circulation around the triple bond occurs in such a way that the protons experience a *diamagnetic shielding* effect. Figure 3.7 shows how this arises, when the axis of the alkyne group lies parallel to the direction of B_0. The cylindrical sheath of π-electrons is induced to circulate around the axis, and the resultant annulus-shaped magnetic field acts in a direction that opposes B_0 in the vicinity of the protons. Higher B_0 values are needed to bring the protons to resonance; therefore acetylenic protons appear at low δ values in the spectrum.

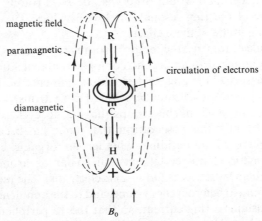

Figure 3.7   *Anisotropic shielding of a proton in an alkyne group.*

Aromatic compounds. In the molecule of benzene (and aromatic compounds in general) π-electrons are delocalised cyclically over the aromatic ring. These loops of electrons are induced to circulate in the presence of the applied field B_0, producing a substantial electric current called the *ring current*. The magnetic field associated with this electric field has the geometry and direction shown in figure 3.8. (An analogy in the macro-world is a ring of copper wire moved into a magnetic field: electric current flows in the wire, and sets up a magnetic field similar in geometry and direction to that shown for benzene in figure 3.8.)

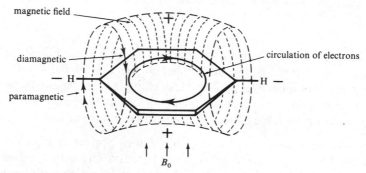

Figure 3.8 *Anisotropic shielding and deshielding associated with the aromatic ring current.*

The induced field is diamagnetic (opposing B_0) in the centre of the ring, but the returning flux outside the ring is paramagnetic (augmenting B_0).

Protons around the periphery of the ring experience a magnetic field greater than B_0, and consequently come to resonance at lower values of B_0 (higher δ values) than would otherwise be so. Protons held above or below the plane of the ring resonate at low δ values. Two examples will illustrate the magnitude of these effects.

In the molecule of toluene, the methyl protons resonate at 2.34 δ, whereas a methyl group attached to an acyclic conjugated alkene appears at 1.95 δ. This is some measure of the greater deshielding influence of the ring current in aromatic compounds (cyclically delocalised π-electrons) compared to the deshielding of conjugated alkene groups (having no cyclic delocalisation). Indeed, so important is this observation, that n.m.r. has become one of the principal criteria used in deciding whether an organic compound has substantial aromatic character (at least in so far as aromatic character relates to cyclic delocalisation of $(4n + 2)$ π-electrons). The method has been applied successfully to heterocyclic systems and to the annulenes; for example [18]annulene sustains a ring current, so that the 12 peripheral protons are deshielded and the 6 internal protons shielded. The outer protons appear at 8.9 δ while the inner protons are *above* TMS at -1.8 δ.

One of the most dramatic observations in n.m.r. work on aromatic systems involves the dimethyl derivative of pyrene (I), in which the methyl groups appear at -4.2 δ, or 4.2 p.p.m. *upfield* of TMS. This shows that the cyclic π-electron system around the periphery of the molecule sustains a substantial ring current, and therefore indicates aromatic character in a nonbenzenoid ring system. The methyl groups are deep in the shielding zone of this ring current, and it is for this reason that they appear at such an extraordinary δ value.

2.34 δ→CH₃

toluene

1.95 δ→CH₃

I
dihydrodimethylpyrene

[18]annulene

pyrene

Alkanes. The equatorial protons in cyclohexane rings come to resonance about 0.5 δ higher than axial protons, and this is attributed to anisotropic deshielding by the σ-electrons in the βγ bonds as shown in figure 3.9. The

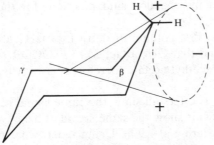

Figure 3.9   *Anisotropic shielding and de-shielding in cyclohexanes.*

effect is small compared with the anisotropic influence of circulating π-electrons.

Simple electronegative (inductive) effects operate only along a chain of atoms, the effect weakening with distance, but magnetic anisotropy operates through space irrespective of whether the influenced group is directly joined to the anisotropic group. For this reason the stereochemistry of molecules must be carefully studied to predict whether magnetically anisotropic groups are likely to have an influence on the chemical shift of apparently distant protons.

3.5 CORRELATION DATA

3.5.1 USE OF CORRELATION TABLES

Although chemical shift data have been rationalised to a large extent using the factors discussed above, much of the application of n.m.r. to organic chemistry is what one might call 'explained empiricism.'

Predicting the n.m.r. spectrum of an organic compound begins with predicting the chemical shift positions for the different hydrogens in the molecule. Figure 3.10 is a useful chart containing approximate proton classifications, and all chemists working with n.m.r. are thoroughly familiar with these allocations.

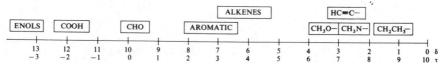

Figure 3.10 *Approximate chemical shift positions for protons in organic molecules.*

Tables 3.4 to 3.9 later in the chapter contain more detailed figures, and a few examples will serve to illustrate the use of these tables.

We shall predict only the approximate chemical shift positions as a drill exercise; single-line signals are not always obtained, and section 3.9 deals with this important complicating feature.

Example 1. Predict the chemical shift positions for the protons in methyl acetate, CH_3COOCH_3.

Table 3.4 shows (first column of data) that CH_3 attached to —COOR appears at 2.0 δ, and that CH_3 attached to —OCOR appears at 3.6 δ. The n.m.r. spectrum of methyl acetate therefore shows two signals, at 2.0 δ and 3.6 δ respectively.

Acetic acid and its esters all show the same signal at ≈ 2.0 δ; all methyl esters of aliphatic acids show the same signal at 3.6 δ (with methyl esters of aromatic acids appearing at 3.9 δ). Using n.m.r. we can therefore diagnose the presence of acetates or methyl esters in an organic molecule with some considerable certainty.

Example 2. Predict the chemical-shift positions for the protons in ethyl acrylate, $CH_2=CHCOOCH_2CH_3$.

Table 3.4 shows (second column of data) that the CH_2 group of an alkyl chain (RCH_2) adjacent to —OCOR appears at 4.1 δ.

The first column of data in the same table shows that the terminal CH_3 group on an alkyl chain (CH_3—R) appears at 0.9 δ, but the —OCOR group nearby, on the β-carbon, has a minor deshielding effect as shown in table 3.5. The value given is +0.4 δ, therefore the CH_3 group appears, not at 0.9 δ but at 1.3 δ.

Ethyl esters, of whatever acid, all have the same chemical shift positions for the CH_2 and CH_3 groups, namely around 4.1 δ and 1.3 δ, respectively.

The δ values for the alkene protons are shown in table 3.7, and three different chemical shift positions are predicted.

alkene δ values   4-nitroanisole

Example 3. Predict the chemical shift positions for the protons in 4-nitroanisole.

Table 3.4 shows that CH_3 attached to —OAr appears at 3.7 δ.

In the aromatic ring of 4-nitroanisole there is considerable symmetry, so that the protons marked a are in identical magnetic environments, and protons b likewise. Chemical shifts in the benzene ring are shown in table 3.9: taking the protons of benzene itself as reference (at 7.27 δ), the —NO₂ group shifts *ortho* protons downfield by 1.0 δ, so that protons a appear at 8.27 δ. The methoxyl group (—OR) moves the protons *ortho* to it upfield by 0.2 δ, so that the b protons appear at 7.07 δ.

(Although the accuracy of the tables does not always justify it, one should also compute the influence of the —NO₂ group on the b protons *meta* to it, and of the MeO— group on the a protons *meta* to it. Thus the a protons are *ortho* to —NO₂ (+1.0 δ) and *meta* to —OR (−0.2 δ), and should appear at (7.27 + 1.0 − 0.2) δ = 8.07 δ. Similarly the b protons are *ortho* to —OR (−0.2 δ) and *meta* to —NO₂ (+0.3 δ), and should appear at (7.27 − 0.2 + 0.3) δ = 7.37 δ.)

In practice one would expect to find the a protons lying somewhere around 8.2 δ, and the b protons between 7.1 and 7.4 δ.

Example 4. Predict the chemical shift positions for the protons in methyl phenoxyacetate, $PhOCH_2COOCH_3$.

As we saw in example 1, methyl esters of acids with an aliphatic residue on the carboxyl group show the methyl signal at 3.6 δ. Table 3.9 shows that the aromatic protons are all moved upfield by the alkoxy group by approximately

0.2 δ (−0.2 for —OR); the aromatic protons should come to resonance around 7.07 δ.

The methylene group, flanked by two electronegative groups, is more difficult to deal with, but Shoolery has computed the relationships listed in table 3.6. For the methylene group in methyl phenoxyacetate, we take the base value of 1.2 δ and add 2.3 δ for —OPh and add 0.7 δ for —COOR, giving 4.2 δ. The Shoolery rules are usually sufficiently accurate for an organic chemist's purpose, particularly for methylene groups, although the error in predicting the δ values for methine protons frequently exceeds 0.5 δ.

3.5.2 INFLUENCE OF RESTRICTED ROTATION

The n.m.r. spectrum of N,N-dimethylformamide, $HCONMe_2$, recorded around room temperature, shows *two* signals for the methyl groups, although it might have been expected that the two methyl groups would be in magnetically equivalent environments. Dimethylformamide is represented by the two resonance forms shown below, and the result of conjugation between the carbonyl group and the nitrogen nonbonding pair is to increase the double-bond character of the C—N bond sufficiently to restrict rotation at room

N,N-dimethylformamide

temperature: one methyl group is *cis* to oxygen, the other is *trans*, and anisotropy of the carbonyl group is sufficient to influence the chemical shift position for the *cis* group. Using a heated probe in the n.m.r. instrument (see supplement 3, page 136 the spectrum of N,N-dimethylformamide can be recorded at high temperature (≈ 130°), and this spectrum shows only one signal for the methyl groups; at elevated temperatures rotation around the C—N bond is so rapid that each methyl group experiences the same *time-averaged* environment.

The four protons in 1,2-dibromoethane ($BrCH_2CH_2Br$) are chemically indistinguishable, and one might also suppose that they are magnetically equivalent; we might predict the same chemical-shift position for all four. The Newman diagrams (opposite) show clearly, however, that protons in the different conformations are not in magnetically equivalent environments. For example, the ringed proton is flanked by H and Br in the first conformer, but by H and H in the second. If the n.m.r. spectrum is recorded at low temperature, the rapid molecular rotations around the C—C bond are 'frozen' sufficiently for these differences to be detected. At room temperature rotation is so rapid that each proton experiences the same time-averaged environment and only one sharp signal appears in the n.m.r. spectrum.

rotational isomers of 1,2-dibromoethane

cyclohexane ring-flips

Similar effects are found in rigid molecules and in structures where steric effects can intervene; examples are cyclic structures (cyclohexane ring-flips, etc.), bridged structures (as in α-pinene, page 90), spirans, etc.

The energy barriers involved in these rotational changes can be measured by n.m.r., using a variable-temperature sample probe, and noting the temperature at which the spectrum changes from that of the mixed conformers to that of the time-averaged situation. Some recent applications of this technique are discussed in supplement 3, page 136. The theory in all cases relates back to $\Delta t \cdot \Delta v \approx 1/2\pi$: we can only resolve the *individual* resonances if the molecule stays in one position longer than Δt; otherwise we see only the time-averaged environment.

3.6 SOLVENTS USED IN N.M.R.

3.6.1 CHOICE OF SOLVENT

To satisfy the condition that nonviscous samples give the sharpest n.m.r. spectra (section 3.2), it is usually necessary to record the spectra of organic compounds in solution; if the compound itself is a nonviscous liquid, the neat liquid can be used. Choice of solvent is not normally difficult, provided a solubility of about 10 per cent is obtainable, but it is clearly an advantage to use aprotic solvents (which do not themselves give an n.m.r. spectrum to superimpose on that of the sample). The following solvents are commonly used, many of which are normal organic solvents in which hydrogen has been replaced by deuterium.

CCl_4	carbon tetrachloride
CS_2	carbon disulphide
$CDCl_3$	deuteriochloroform
C_6D_6	hexadeuteriobenzene
D_2O	deuterium oxide (heavy water)

$(CD_3)_2SO$ hexadeuteriodimethylsulphoxide
$(CD_3)_2CO$ hexadeuterioacetone
$(CCl_3)_2CO$ hexachloroacetone

A more comprehensive list, including more precise data, is given in table 3.11.

For deuteriated solvents the isotopic purity should ideally be as high as possible, but since greater isotopic purity means greater cost, most users compromise their ideals. Isotopic purity approaching 100 per cent can often be obtained but, for example, a small $CHCl_3$ peak at 7.3 δ in 99 per cent $CDCl_3$ will cause no difficulty or ambiguity unless accurate integrals for aromatic protons (also around 7.3 δ) are demanded.

3.6.2   SOLVENT SHIFTS—CONCENTRATION AND TEMPERATURE EFFECTS

The solvents listed above vary considerably in their polarity and magnetic susceptibility. Not surprisingly, the n.m.r. spectrum of a compound dissolved in one solvent may be slightly different from that measured in a more polar solvent, and it is important in all n.m.r. work to quote the solvent used. The n.m.r. signals for protons attached to carbon are, in general, shifted only slightly by changing solvent, except where significant bonding or dipole–dipole interaction might arise: the n.m.r. spectrum for chloroform dissolved in cyclohexane appears at 7.3 δ, but in benzene solution the signal is moved upfield by the exceptionally large amount of 1.56 δ (to 6.74 δ). Benzene is behaving as a Lewis base to chloroform, and considerable charge transfer is responsible for altering the electron density around the chloroform proton, with concomitant upfield shift in the signal.

In contrast, NH, SH, and particularly OH, protons all have their n.m.r. signals substantially moved on changing to solvents of differing polarity. This effect is largely associated with hydrogen bonding, and it is noted even when different concentrations are used in the same solvent.

At low concentrations, intermolecular hydrogen bonding is diminished in simple OH, NH, and SH compounds: since hydrogen bonding involves electron-cloud transfer from the hydrogen atoms to a neighbouring electronegative atom (O, N, or S), the hydrogen experiences a net deshielding effect when hydrogen bonding is strong, and is less deshielded when hydrogen bonding is diminished. Thus, at high concentrations (strong hydrogen bonding, strong deshielding) OH, NH and SH protons appear at higher δ than in dilute solutions.

Intermolecular H-bonding: δ values lowered with increased dilution or temperature

Table 3.8 lists the chemical shift positions for protons subject to hydrogen bonding, and it can be seen that the range within which they come to resonance is wide (0.5–4.0 δ for simple alcohols).

Increased temperature also reduces intermolecular hydrogen bonding, so the resonance positions for these protons are temperature dependent (higher temperatures mean lower δ values).

Intramolecular hydrogen bonding is unchanged by dilution and the n.m.r. spectrum from such systems is virtually unaltered by varying concentration or temperature. Salicylates and enols of β-dicarbonyl compounds are examples of such systems: chelates, such as the salicylates, show the OH resonance at very high δ (10–12 δ), and enol OH appears even higher (11–16 δ).

Carboxylic acids are a special case of hydrogen bonding because of their stable dimeric association, which persists even in very dilute solution; carboxylic OH appears between 10 and 13 δ, usually nearer 13 δ.

salicylates   enol of β-diketone   carboxylic acid dimer
Intramolecular H-bonding: δ values unchanged by concentration or temperature changes

3.7 INTEGRALS

We mentioned very briefly in the introduction to this chapter that the area under each n.m.r. signal in the spectrum is proportional to the number of hydrogen atoms in that environment. The two peaks in the toluene spectrum (figure 3.1) have relative areas 5:3 corresponding to C_6H_5 and CH_3, respectively.

Measurement of the peak areas is carried out automatically on the n.m.r. spectrometer by integration of each signal, and the integral value is indicated on the spectrum in the form of a continuous line in which steps appear as each signal is measured: step height is proportional to peak area. These *integral traces* are shown on the spectra in problems 6.2(i) to 6.2(v) (pages 228–233).

Note that only *relative* peak areas are recorded, and thus only the *ratio* of protons in each environment is given. The step heights should be measured to the nearest 0.5 mm, and by proportionality the best-fit *integer ratio* calculated.

Accuracy. The accuracy of the integrator on the instrument will be specified by the manufacturer, and is commonly within 2 per cent for several consecutive scans of a standard spectrum (for example that of ethylbenzene), but a number of factors militates against this accuracy applying to the spectra of specific compounds. The net absorption of radiofrequency energy in the n.m.r. experiment depends on the relaxation time T_1, on the rate of scan and on the intensity of the radiofrequency source. Not all protons have the same relaxation times (see section 3.2), and slight deviations from exact integer ratios can be found in the signal intensities. If the irradiating frequency is too intense in relation to the scanning rate, saturation of the signals may arise and lead to low integral values. A spectrum with a noisy baseline, perhaps because a weak signal has to be greatly amplified, may not give an accurate integral trace, since noise in the baseline will be integrated along with the resonance signals. Broad peaks tend also to give less accurate integrals than sharp peaks.

3.8 SPIN COUPLING—SPIN–SPIN SPLITTING

3.8.1 THE SPLITTING OF n.m.r. SIGNALS
The n.m.r. spectrum of *trans*-cinnamic acid is reproduced in figure 3.11.

The aromatic protons, five in number, give rise to the large peak at 7.4 δ, and the carboxyl proton is off-scale at 13.0 δ; both of these signals we might have predicted from the discussions in sections 3.4 and 3.5.

What we would not have predicted from chemical-shift data alone is that proton H_A appears as *two* lines on the spectrum (centred on 6.5 δ), and proton H_X appears as *two* lines (centred on 7.8 δ). We say that each signal is *split into a doublet*. Note that the separation between the two H_A lines is the same as the separation between the two H_X lines.

Look now at the spectrum of 1,1,2-trichloroethane (figure 3.12). The signal from proton H_A appears as a *triplet*, while that from protons H_X is a *doublet*.

The number of lines (multiplicity) observed in the n.m.r. signal for a group

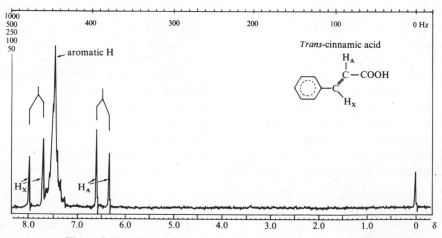

Figure 3.11   *N.M.R. spectrum of trans-cinnamic acid.* (CDCl$_3$).

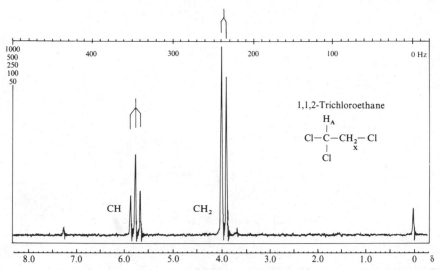

Figure 3.12 *N.M.R. spectrum of 1,1,2-trichloroethane.* (CCl$_4$).

of protons is *not* related to the number of protons *in that group*; the multiplicity of lines is related to the number of protons in *neighbouring groups*.

For example, protons H$_X$ in figure 3.12 have only *one* neighbouring proton, and H$_X$ appears as a two-line signal (doublet); proton H$_A$ has *two* neighbours and the signal is split into three lines (triplet).

(*n* + 1) *rule*. The simple rule is: to find the multiplicity of the signal from a group of protons, count the number of neighbours (*n*) and add 1. (Exceptions to the rule are discussed in section 3.10.)

Splitting of the spectral lines arises because of a *coupling* interaction between neighbour protons, and is related to the number of possible spin orientations that these neighbours can adopt. The phenomenon is called either *spin–spin splitting* or *spin coupling*.

3.8.2 THEORY OF SPIN–SPIN SPLITTING

The diagram in figure 3.13 represents two vicinal protons similar to the alkene protons in cinnamic acid, H$_A$ and H$_X$. These protons, having different chemical and magnetic environments, come to resonance at different positions in the n.m.r. spectrum; they do not give rise to single peaks (singlets) but doublets. The separation between the lines of each doublet is equal: this spacing is called the *coupling constant, J*.

Why is the signal for proton A split into a doublet? A simplistic explanation is that the resonance position for A depends on its total magnetic environment; part of its magnetic environment is the nearby proton X, which is itself magnetic, and proton X can either have its nuclear magnet *aligned with* proton A or *opposed to* proton A. Thus proton X can either *increase* the net

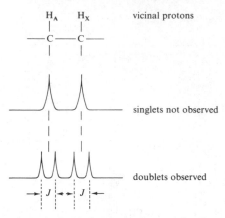

Figure 3.13 *Splitting in the signals of two vicinal protons.*

magnetic field experienced by A (X aligned) or *decrease* it (X opposed); in fact it does both. The two spin orientations of X create two different magnetic fields around proton A. Therefore proton A comes to resonance, not once, but twice, and proton A gives rise to a doublet.

Similarly, proton A is a magnet having two spin orientations with respect to A, and A creates two magnetic fields around X. Proton X comes to resonance twice in the n.m.r. spectrum.

This mutual magnetic influence between protons A and X is not transmitted through space, but via the electrons in the intervening bonds. The nuclear spin of A couples with the electron spin of the C—H_A bonding electrons; these in turn couple with the C—C bonding electrons and then with the C—H_X bonding electrons. The coupling is eventually transmitted to the spin of the H_X nucleus. This *electron-coupled* spin interaction operates strongly through one bond or two bonds, less strongly through three bonds, and, except in unusual cases, rather weakly through four or more bonds. This point is more rigorously developed in the following section.

We can represent the possible spin orientations of coupling protons as in figure 3.14. Proton A can 'see' proton X as aligned (parallel ↑) or opposed (antiparallel ↓); these two spin orientations correspond effectively to two different magnetic fields. Therefore proton A comes to resonance twice. The same argument explains why proton X appears as a doublet.

The A and X protons of cinnamic acid give rise to this characteristic pair of doublets, caused by two protons undergoing spin coupling: such a spectrum is called an AX spectrum. Since the probability of the two spin orientations of A and X arising is equal in molecules throughout the sample, the two lines in each doublet are of equal intensity. (See, however, section 3.10.)

Figure 3.15 represents the coupling that arises in the triplet signal in the n.m.r. spectrum of 1,1,2-trichloroethane.

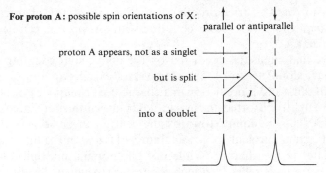

Figure 3.14 *Representation of spin coupling between a proton A and one neighbouring proton X.*

When proton A 'sees' the two neighbouring protons X and X', then A can 'see' *three* different possible combinations of spin: (1) the nuclear spins of X and X' can both be parallel to A (↑↑); (2) both can be antiparallel to A (↓↓); (3) one can be parallel and the other antiparallel, and this can arise in two ways—X parallel with X' antiparallel (↑↓) or X antiparallel with X' parallel (↓↑). Three distinct energy situations, (1), (2) and (3), are created, and therefore proton A gives rise to a triplet. The probability of the first two energy states arising is equal, but since the third state can arise in two different ways it is twice as likely to arise; the intensity of the signal associated with this state is twice that of the lines associated with the first two states, and we see in the spectrum of 1,1,2-trichloroethane the relative line intensities in the triplet are 1:2:1.

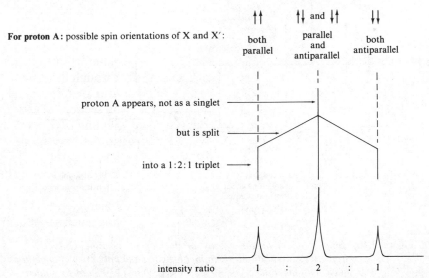

Figure 3.15 *Representation of spin coupling between a proton A and two neighbour protons X and X'.*

The spectrum of 1,1,2-trichloroethane, consisting of the characteristic doublet and triplet for two protons coupling with one proton, can be called an AX_2 spectrum.

The relative line intensities predicted by the above spin coupling mechanism are 1:1 in doublets, and 1:2:1 in triplets. Real spectra almost always depart from this first-order prediction in a characteristic manner: the doublet and triplet are slightly distorted, the inner lines being a little more intense than the outer lines. This 'humping' towards the centre can be seen in the AX coupling in the cinnamic acid spectrum, figure 3.11, and in the other spectra reproduced later in the chapter. While not all coupling multiplets 'hump' towards the centre, it is a reliable enough occurrence to help in deciding which are the coupling multiplets in complex spectra.

3.8.3 MAGNITUDE OF THE COUPLING—COUPLING CONSTANTS, J

The coupling constant J is a measure of the interaction between nuclei, and we have stated earlier that the interaction is transmitted through the intervening electrons: for two nuclei, there are four energy levels involved in the n.m.r. transitions, and their relative positions are governed by the internuclear spin coupling as shown in figure 3.16. When there is zero coupling

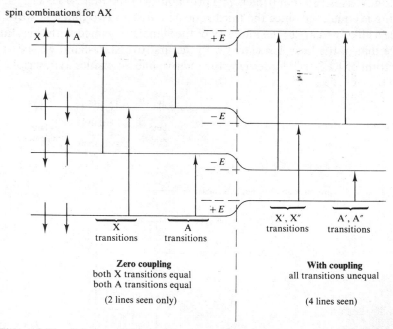

Figure 3.16 *Spin coupling as the energy of nuclear interaction. Only those transitions are allowed which involve one (and only one) change in nuclear spin; thus ↓↑→↑↑ is allowed, but ↓↑→↑↓ is forbidden. Example shown is for J positive: for negative J the ±E changes in energy levels are reversed.*

between A and X, both X transitions are equal as are both A transitions: each nucleus gives rise to one line absorption.

When coupling takes place, we can suppose that the energy levels are altered by $\pm E$, the energy of interaction, then the transitions no longer remain equal: the X transition splits into an X' line (transition energy $X + 2E$) and an X" line $(X - 2E)$. The A transition is likewise split into A' $(A + 2E)$ and A" $(A - 2E)$. The spacings between X' and X" and A' and A" are equal, and have magnitude $4E$: J, the coupling constant, is therefore equal to $4E$, and is a measure of nuclear interaction, *which is wholly independent of any external magnetic field*. The units of J are energy units, usually Hz; and it is easy to see that J can have sign as well as magnitude. The sign of J is hardly ever of importance to organic chemical applications of n.m.r., but it is discussed again in supplement 3 in connection with spin tickling and I.N.D.O.R. In practical terms, it is simpler to interpret signal multiplicity in more extensive coupling systems by utilising the $(n + 1)$ rule and the concept of x nuclei 'seeing' y neighbouring nuclei, etc.

3.8.4   MORE COMPLEX SPIN–SPIN SPLITTING SYSTEMS

The two examples of spin coupling above, the AX and AX_2 cases, show virtually undistorted multiplets, with the multiplicity following the $(n + 1)$ rule, and signal intensities almost exactly 1:1 and 1:2:1. Such spectra are described as *first-order* spectra. We can extend this first-order treatment successfully to more complex systems, before considering non-first-order spectra (in section 3.10).

The spectrum of 2-chloropropionic acid (figure 3.17) contains a doublet and a quartet (J, 8 Hz), corresponding to the coupling of one proton with three neighbours; this is an AX_3 spectrum.

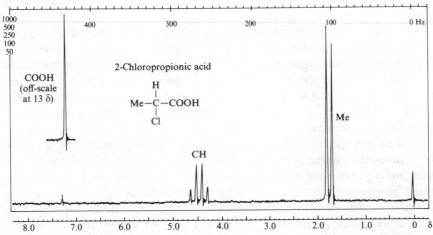

Figure 3.17   *N.M.R. spectrum of 2-chloropropionic acid. (CDCl$_3$. Note CHCl$_3$ impurity peak at 7.3 δ.)*

The methyl protons have one neighbour, and therefore appear as a doublet.

The methine proton has three neighbours, on the methyl group, and therefore $(n + 1)$ is 4. For this methine proton, we must consider the various ways in which the spin orientations of the methyl protons can be grouped together, and we find that four arrangements are possible: (1) the methine proton can 'see' all three spins of the methyl protons parallel ($\uparrow\uparrow\uparrow$); (2) alternatively, two spins may be parallel with one antiparallel (and there are three ways in which this can arise —$\uparrow\uparrow\downarrow$ or $\uparrow\downarrow\uparrow$ or $\downarrow\uparrow\uparrow$); (3) then two spins can be antiparallel with one parallel (arising in three ways—$\downarrow\downarrow\uparrow$ or $\downarrow\uparrow\downarrow$ or $\uparrow\downarrow\downarrow$); (4) lastly all three spins can be antiparallel ($\downarrow\downarrow\downarrow$).

Four different energy states are produced; therefore the methine proton comes to resonance four times, appearing as a quartet. The relative probabilities of these states arising are in the ratio 1:3:3:1, and the line intensities in the quartet have the same ratio.

One can go further and predict the theoretical line intensities for quintets, sextets, etc., and find that the ratios are the same as the coefficients in the binomial expansion. Pascal's famous triangle serves to remind.

			1				singlet
		1		1			doublet
	1		2		1		triplet
1		3		3		1	quartet
1	4		6		4	1	quintet
1	5	10	10	5	1		sextet

The outer lines in substantial multiplets are of such low intensity that they may be all but unobservable, unless that part of the spectrum is re-run on expanded scale.

The spectrum of ethyl bromide (figure 3.18) is an A_2X_3 case, and the triplet and quartet that this engenders is one of the easiest of systems to identify in an n.m.r. spectrum. All isolated ethyl groups produce a similar spectrum, the chemical shift positions of the CH_2 protons being dependent on substituent (see example 2, page 95). The presence of such ethyl groups, as in ethyl esters, ethyl ketones, ethyl ethers, and in ethanol, etc., are quite unequivocally identifiable from the n.m.r. spectrum.

In the n.m.r. spectrum of 1-nitropropane (figure 3.19) there are three groups of protons, each group coupling with its near neighbours, so that the central methylene group couples both with the methyl protons and with the terminal methylene protons. The methyl group appears as a triplet, and the terminal methylene group also appears as a triplet, since both couple with the central methylene group: $(n + 1) = (2 + 1) = 3$. (The methyl protons show no coupling with the terminal methylene protons since coupling over four σ-bonds is rarely observed; but see virtual coupling, page 121).

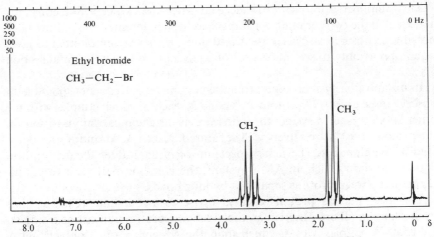

Figure 3.18   *N.M.R. spectrum of ethyl bromide.* (CDCl$_3$)

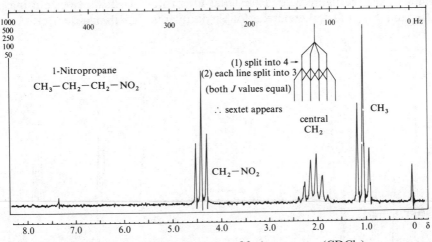

Figure 3.19   *N.M.R. spectrum of 1-nitropropane.* (CDCl$_3$)

The central methylene group can be dealt with by considering the successive coupling, first with the methyl group, and then with the terminal methylene protons. The methyl protons split the central methylene signal into a quartet; the two terminal methylene protons should now split *each line of this quartet* into a triplet, giving twelve lines in all. In the spectrum only six lines are observed, showing that considerable overlapping of the predicted twelve lines has taken place. In fact the two coupling constants involved (J_{CH_3, CH_2} and J_{CH_2, CH_2}) are equal, and an easier way to consider the central methylene group is to add the *total* number of neighbours with which it couples (CH_3 and CH_2) and then apply the ($n + 1$) rule. There are five coupling neighbours: therefore the multiplicity is 6.

This simplified approach only succeeds when the two coupling constants are equal. If the two coupling constants had been different, it might have been possible to observe all twelve predicted lines; the spectrum of ethanol gives us an opportunity to see this degree of coupling; see figure 3.22 and section 3.8.5.

In unsaturated systems (and aromatic systems) it is frequently possible to observe three groups of protons A, M and X, each of which couples with the other two. For such a system to be first order, the chemical shift positions of the protons must be relatively well separated, just as A, M and X are separated in the alphabet. The n.m.r. spectrum of furan-2-aldehyde (furfural) in figure 3.20 shows such an AMX system; the coupling pattern involves interaction between protons separated by four bonds.

The three nuclear protons each give rise to a four-line signal, so that twelve lines in all are observable; the aldehydic proton appears as a singlet.

Proton A couples with X, which splits the A signal into a doublet; but A also couples with M, so that *each line* of A is further split in two, giving four lines in all; see figure 3.20. The signal for proton A shows two splittings, J_{AM} and J_{AX}, and is therefore a double doublet (rather than a quartet).

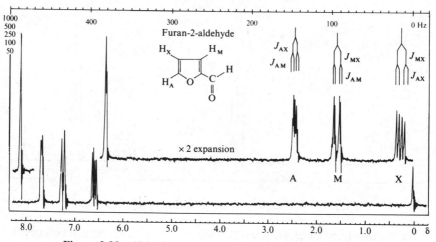

Figure 3.20   *N.M.R. spectrum of furan-2-aldehyde.* (CDCl$_3$).

Similarly the M signal is split into two by coupling with X, and each line is further split into two by coupling with A; two coupling constants are again seen, J_{AM} and J_{MX}, and M appears as a double doublet.

Lastly the X signal is split into a double doublet by two successive couplings, J_{MX} and J_{AX}.

A full analysis of an AMX spectrum involves identifying all three J values, J_{AM}, J_{AX} and J_{MX}. Note that each J value appears in two different multiplets, and that each multiplet contains two different J values.

Origin chemical shift positions of A, M and X

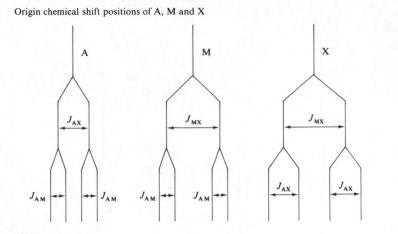

Final appearance of the spectrum

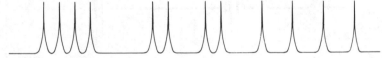

Figure 3.21 *Coupling constants in an* AMX *system.*

The n.m.r. spectrum of many vinyl compounds (for example vinyl acetate) show AMX coupling systems.

Figure 3.21 shows the AMX coupling system in schematic form.

3.8.5 PROTON-EXCHANGE REACTIONS

If the n.m.r. spectrum of ethanol is recorded on a typical commercial sample and then compared with the spectrum of a high-purity sample, we become faced with yet another problem; figure 3.22 shows these two spectra. The spectrum of the commercial-grade sample is easily explained using the multiplicity predictions for an A_2X_3 case as for ethyl bromide (figure 3.18), but clearly the OH proton is not involved in coupling with the CH_2 group. The spectrum for the pure sample does show this coupling, and we can explain the multiplicity using the arguments developed for the 1-nitropropane spectrum (figure 3.19).

The CH_3 group is a triplet because of coupling to CH_2. The OH signal is a triplet because of coupling to CH_2. The CH_2 is split into a quartet by the CH_3 group, and each line is further split into two by the OH proton. There are two different coupling constants involved (8 Hz and 6 Hz) so all eight predicted lines are reasonably clear.

Why is the OH coupling not observed in the spectrum recorded on the 'normal' commercial sample of ethanol?

Exchange of the OH protons among ethanol molecules is normally so rapid that one particular proton does not reside for a sufficiently long time on

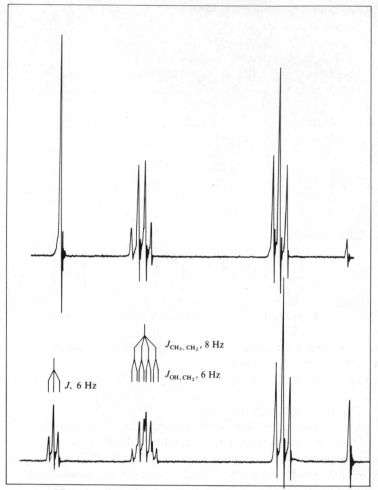

Figure 3.22   *N.M.R. spectrum of ethanol. Top: sample with acidic impurities. Bottom: pure sample, showing* OH–CH$_2$ *coupling (recorded at 100 MHz).*

a particular oxygen atom for the nuclear coupling to be observed

$$R—O—H + R—O—H^* \rightleftharpoons R—O—H^* + R—O—H$$

We saw earlier (page 82) that $\Delta t \cdot \Delta v \approx 1/2\pi$. Here, Δt is the time needed to resolve accurately the multiplicity in the CH$_2$ and OH groups brought about by their coupling, and Δv is the coupling constant. Provided the *residence time* of a particular proton on oxygen is sufficiently long (longer than Δt) we can record the coupling. If there is rapid proton exchange the residence time will be shorter than Δt, and the coupling will not be resolved. The rate of exchange is related to the coupling constant (6 Hz in this case) and if we are unable to resolve the coupling, the rate of exchange must be greater than $6\,s^{-1}$. The exchange is acid-catalysed and base-catalysed, and only in samples that are acid and base free is the residence time sufficiently long for the coupling to be observed between OH and CH_2.

Rapid proton exchange occurs in carboxylic acids, phenols, amines, amides and thiols etc. (see table 3.8), so that in general no coupling is observable between the protons on these functions and their neighbours.

In the case of alcohols, it is relatively easy to see the OH coupling if the sample is pure, or if the spectrum is recorded with a low-temperature sample (when the exchange is retarded), or if the sample is dissolved in a highly polar solvent such as dimethylsulphoxide, Me_2SO, when strong solvation presumably stabilises individual molecules and reduces the exchange.

An important corollary of this process is deuterium exchange, which is dealt with in section 3.9.5.

3.9 FACTORS INFLUENCING THE COUPLING CONSTANT J

3.9.1 GENERAL FEATURES

As was pointed out earlier, the coupling constant J can have positive or negative values; initial interpretation of the spectra can profitably omit this factor, since the sign of J cannot be directly extracted by observation.

Table 3.10 lists the commonest proton–proton coupling constants found in organic molecules, and some general points are worth highlighting before beginning a general discussion.

Geminal coupling, involving protons separated by only two bonds, is strong, being typically 10–18 Hz, but it will only be observed where the *gem* protons have different chemical shift positions as discussed in sections 3.9.2 and 3.10.

Vicinal coupling (three bonds separating the protons) varies from 0–12 Hz in rigid systems, but in freely rotating carbon chains (alkyl groups) it is usually around 8 Hz.

Trans coupling in alkene groups (J, 11–19 Hz) is stronger than *cis* coupling (J, 5–14 Hz).

Aromatic coupling depends on whether the coupling protons are *ortho*, *meta* or *para* to each other, and in simple cases the coupling constant is definitive in deciding the orientation; thus J_{ortho}, 7–10 Hz; J_{meta}, 2–3 Hz; J_{para}, 0–1 Hz.

Allylic coupling as in allyl chloride is the most likely four-bond coupling to be met in nonaromatic molecules and is very small (J, 0–2 Hz). The analogous coupling in aromatic systems (for example between the methyl protons and the *ortho* protons in the ring) is not normally large enough to be measured, although it has been observed in certain polynuclear aromatic hydrocarbons such as 2-methylpyrene in which considerable double-bond character exists in the intervening aromatic C=C bond.

CH$_2$=CHCH$_2$Cl
allyl chloride

2-methylpyrene

$$\begin{array}{ccc} H_A & & H_X \\ | & & | \\ -C{=}C{-}C{-} \\ & & | \end{array}$$

J_{AX}, allylic coupling, 0–2 Hz

Other magnetic nuclei present in the molecule (^{14}N, ^{19}F, etc.) may increase the complexity of proton spectra (see section 3.9.4), while substitution of deuterium for hydrogen may lead to simplification (see sections 3.9.4 and 3.9.5).

3.9.2 FACTORS INFLUENCING GEMINAL COUPLING
The electronegativity of an attached substituent alters the value of *gem* coupling, but not always predictably. In groups such as —CH$_2$—X, the *gem* coupling will range from 12–9 Hz as the electronegativity of X is increased. These couplings cannot be measured directly because the two protons have identical δ values, but in the derived —CHD—X the *gem* coupling between H and D (H—C—D) can be measured; $J_{H,H}$ can then be calculated from the equation $J_{H,H} = 6.53 J_{H,D}$ (see section 3.9.4).

The magnitude of J_{gem} also varies with the H—$\widehat{C}$—H bond angle, being of greatest magnitude (10–14 Hz) in the strain-free cyclohexanes and cyclo-pentanes. With increasing angular strain the value of J_{gem} drops, being 8–14 Hz in cyclobutanes and 4–9 Hz in cyclopropanes.

3.9.3 FACTORS INFLUENCING VICINAL COUPLING
The electronegativity of attached substituents alters the value of vicinal coupling, as it does of geminal. In qualitative terms, the more electronegative the substituent the smaller the value of J_{vic}, so that in unhindered ethanes the value is ≈ 8 Hz, and in halogenoethanes it is lowered to 6–7 Hz. Where there is restricted rotation, the angle subtended by the electronegative sub-stituent at the C—C bond also has an effect on J_{vic}, and other constraints which alter the angles H—$\widehat{C}$—C and C—$\widehat{C}$—H, particularly the presence of small rings, will influence J_{vic}.

The factor that is ostensibly most easy to predict in its influence on J_{vic} is the dihedral angle ϕ between the two vicinal C—H bonds; the equations due to Karplus give frequent agreement with the observed values.

Karplus's equations

$$\phi \text{ between } 0° \text{ and } 90°: J_{vic} = 8.5 \cos^2 \phi - 0.28$$
$$\phi \text{ between } 90° \text{ and } 100°: J_{vic} = 9.5 \cos^2 \phi - 0.28$$

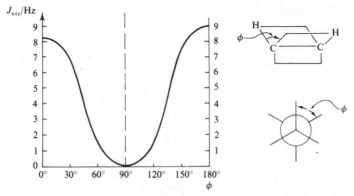

Figure 3.23   *Variation of vicinal coupling constant, J_{vic}, with dihedral angle ϕ. (graphical presentation of Karplus's equations)*

It is more convenient to express this graphically (see figure 3.23), and indeed the reliability of the method is not sufficiently high to justify accurate calculations.

To summarise the Karplus rules, the largest vicinal couplings arise with protons in the *trans* coplanar positions ($\phi = 180°$). Vicinal couplings for *cis* coplanar protons are almost as large ($\phi = 0°$). Very small couplings arise between protons at 90° to each other.

As an example of the successful application of these rules we can consider the protons in chair cyclohexanes; diaxial protons have coupling constants around 10–13 Hz (somewhat larger than predicted) and this collates with their 180° orientation; diequatorial protons, or those with axial/equatorial relationship have coupling constants around 2–5 Hz, corresponding to about 60° orientation.

3.9.4   HETERONUCLEAR COUPLING

The presence in an organic molecule of magnetic nuclei other than hydrogen can introduce additional complications into the spectrum, since these nuclei will take up spin orientations with respect to the applied field and may cause spin–spin splitting of the proton signals in the same way as neighbour protons do.

As a simple example, the spectrum of 2,2,2-trifluoroethanol is shown in figure 3.24. The CH_2 group does not appear as a singlet, but couples with the three vicinal ^{19}F atoms; since ^{19}F has $I = \frac{1}{2}$, each fluorine nucleus can take up two spin orientations in exactly the same manner as hydrogen.

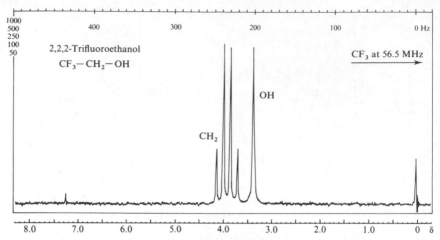

Figure 3.24 *N.M.R. spectrum of 2,2,2-trifluoroethanol* ($CDCl_3$). *The* CH_2 *group is split into a quartet by the three neighbour* ^{19}F *atoms.*

The splitting of the CH_2 group by CF_3 in CF_3CH_2OH is analogous to the splitting of the CH_2 group by CH_3 in CH_3CH_2OH; the coupling constants are about the same (8 and 11 Hz, respectively, for H–H and H–F coupling).

The signal for ^{19}F itself is off scale, about 60 000 ppm upfield from TMS; if we wish to observe this at the same field strength (1.4 T) we need a radio-frequency source at 56.5 MHz, the precessional frequency for ^{19}F. Couplings with ^{19}F, ^{31}P and ^{13}C are dealt with in more detail in supplement 3.

Deuterium (^2H) and nitrogen (^{14}N) have $I = 1$, and therefore have three spin orientations ($2I + 1$) in an applied field; when ^1H couples with these nuclei the ($n + 1$) rule has to be modified.

Deuterium coupling. The three spin states for ^2H correspond to -1, 0, and $+1$, *all equally populated.* A proton coupling with one deuterium nucleus will 'see' three spin orientations for the deuterium nucleus, and therefore experience three different magnetic fields, and come to resonance three times. In summary, proton coupled with deuteron appears as a triplet; since the three spin states of deuterium are equally populated, the probabilities for the three energy states are equal, so the line intensities are 1:1:1(and not 1:2:1 as in AX_2 proton triplets).

If we record the ^1H n.m.r. spectrum of a $\rangle CHD$ group, the proton signal will appear as a 1:1:1 triplet. (We can only see the deuterium n.m.r. signal itself if we substitute an appropriate radiofrequency probe at 9.2 MHz, the precession frequency of ^2H at 1.4 T: the deuterium signal will be a doublet, because of coupling to ^1H.)

Nitrogen coupling. The ^1H spectrum for N—H groups should show splitting similar to the deuterium case, since ^{14}N also has three spin orientations, equally populated. As we saw in section 3.2, page 82, we must also recognise the efficient spin–lattice relaxation for protons attached to ^{14}N. Because of its electric quadrupole, nitrogen can efficiently provide the vectors necessary for excited protons to relax. Since T_1 is short, the ^1H signal is usually broadened, and the ^{14}N spin splitting of the proton resonance is not resolved. (As in the cases of fluorine and deuterium, and appropriate radiofrequency probe—4.3 MHz at 1.4 T—is needed to observe the ^{14}N n.m.r. itself.)

3.9.5 DEUTERIUM EXCHANGE

If deuterium oxide D_2O is used as solvent for n.m.r. work, the D_2O exchanges with labile protons such as OH, NH and SH. The mechanism is the same as for proton exchange discussed in section 3.8.5, page 109. In effect, because of the rapidity of the exchange, ROH becomes ROD, RCOOH becomes RCOOD, $RCONH_2$ becomes $RCOND_2$, etc.

$$ROH + D—O—D\,(D_2O) \rightleftharpoons ROD + H—O—D$$

Peaks previously observed for the labile protons disappear (or are diminished) and a peak corresponding to H—O—D appears around 5 δ.

This technique of *deuteration* is widely used to detect the presence of OH groups, etc., and is easily carried out. The n.m.r. spectrum can be recorded conventionally in a solvent other than D_2O, and then a few drops of D_2O are shaken with the sample and the spectrum is re-run. Peak areas diminish for OH, NH signals, etc.

The method can be extended to detect reactive methylene groups, such as those flanked by carbonyl. The spectrum is recorded normally, and then a small amount of D_2O and sodium hydroxide are added to the sample tube; base-catalysed deuterium exchange occurs only on reactive sites, for example

When the spectrum is re-run, signals from protons previously on these sites disappear.

3.10 NON-FIRST-ORDER SPECTRA

In general, first-order spectra will only arise if the separation between multiplets (the chemical shift difference between signals) is much larger than the coupling constant J; if $\Delta\delta \geqslant 6\,J$ then fairly unperturbed spectra will arise.

For example, in the spectrum of 1,1,2-trichloroethane, figure 3.12, the coupling constant J is 8 Hz, while the difference between the chemical shift positions of CH and CH_2 is $(5.7-3.9)$ $\delta = 1.8$ δ, which corresponds to 108 Hz on a 60 MHz spectrum.

If the signals from coupling protons are closer together on the spectrum, and the chemical shift difference is small, distortion of the signals arises. Figure 3.25 shows a simple example of this in the AX case (for example cinnamic acid, figure 3.11). As the signals move together, the inner peaks become even larger at the expense of the outer peaks, and the positions of the lines also change: the origin chemical shift positions are no longer found at the mid-points of the doublets, but lie approximately at the 'centre of gravity' of the doublets. Such spectra are usually called AB spectra, indicating that the chemical shift values are closer than in AX cases.

In the ultimate case of two protons having the same chemical shift, so

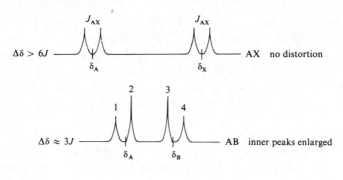

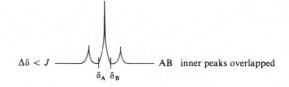

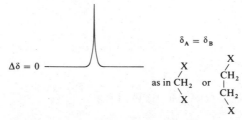

Figure 3.25   *Effect of ratio $\Delta\delta : J$ on doublet appearance. For AB cases, the origin chemical shift positions can be calculated from $\nu_A - \nu_B = [(\nu_1 - \nu_4)(\nu_2 - \nu_3)]^{1/2}$.*

that $\Delta\delta = 0$ (for example the *gem* protons of most unhindered CH_2 groups) the interaction between spins, although present at the nuclear level, is not observable in the n.m.r. spectrum and a singlet peak is produced.

As the chemical shift values of coupling protons approach equality, the energy levels involved in the transitions among spin states become closer: new transitions frequently become possible, and additional lines may appear on the spectrum. Such interactions also give rise to changes in the expected Boltzmann distributions among spin states, and the predicted first-order line intensities are distorted as in the AB diagram above. Population changes in the spin states can also be brought about deliberately, as discussed in section 3S.1 (double resonance techniques).

In the AX_2 case of 1,1,2-trichloroethane (figure 3.12) only five lines appear, as predicted by first-order $(n + 1)$ rules. If the chemical shift values of A and X are closer together, the spectrum is non-first-order (and should be called AB_2); altogether nine lines may appear, the ninth line often being vestigial. The precise appearance of the spectrum depends on the ratio $\Delta\delta : J$, as shown in figure 3.26. Origin chemical shift positions are no longer easily extracted with exactitude, but they again lie near enough to the centres of gravity of the multiplets for most structural organic analyses. Where eight or nine lines can clearly be seen, the origin chemical shift positions are as indicated in figure 3.26; J_{AB} in these circumstances has to be calculated from the relationship

$$J_{AB} = 1/3(v_4 + v_8 - v_1 - v_6)$$

where v_1, etc. are the chemical shifts in Hz from TMS.

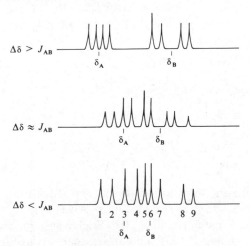

Figure 3.26 *Some possible appearances of AB_2 spectra. Where nine lines can clearly be seen, δ_A lies at line 3, and δ_B at the mid-point of lines 5 and 7.*

For three coupling environments, the first-order appearance is AMX (see above). If two of the origin chemical shift positions are closer than in a true AMX case, then the ABX system results, which in many cases, like AMX, exhibits twelve lines. Chemical shift positions and coupling constants cannot easily be extracted from the spectrum directly, but sufficient accuracy for structural work can be achieved by treating a clear twelve-line ABX spectrum as an AMX case.

If all three coupling environments are close together in chemical shift terms, ABC systems arise, in which case up to fifteen lines can be observed. No attempt should be made to extract coupling constant data from such spectra, since none of the line spacings represents a true J value: chemical shift data should be evaluated on an approximate basis only. Many ABC spectra have been fully analysed by calculation, and specialist texts (for example Wiberg and Nist, listed in Further Reading) should be consulted if detailed δ and J values are required.

Aromatic systems. The majority of aromatic coupling systems are non-first-order, and many examples can be seen in perusing spectra catalogues (for example the Varian catalogues). It nevertheless serves a useful purpose to describe the most frequently encountered cases, beginning with single-line spectra. The characteristic appearance of a selection of simple systems is shown in figure 3.27. Scale expansion will show further small splittings.

Single-line aromatic spectra are produced by several monosubstituted benzenoid derivatives, provided the substituent has no strong shielding or deshielding effect (for example toluene, I). Compounds with identical *para* substituents, whatever their electronegativity, give single-line spectra because of the molecular symmetry; all four protons in the ring are magnetically equivalent (for example *p*-dinitrobenzene, II).

I   II   III   IV   V

Unsymmetrical *para* substitution, if the two substituents have different shielding influences, can give rise to almost AB simplicity for the aromatic protons (for example *p*-chloroaniline, III). Such a system is easily recognised in the spectrum (see figure 3.27) and can be analysed as for the AB case to which it approximates; but because each proton couples with the protons *ortho* and *para* to it, additional transitions become possible. Additional lines appear, characteristically within each doublet (see figure 3.27) and the system is more correctly described as the non-first-order AA′BB′ system.

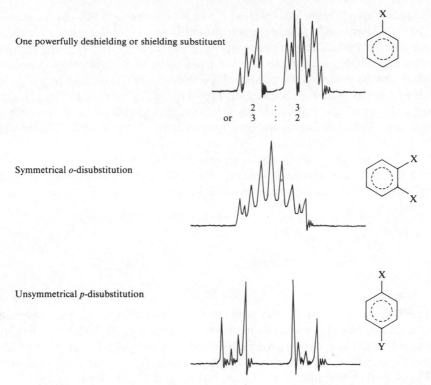

One powerfully deshielding or shielding substituent

or $\begin{matrix} 2 & : & 3 \\ 3 & : & 2 \end{matrix}$

Symmetrical *o*-disubstitution

Unsymmetrical *p*-disubstitution

Figure 3.27   *The appearance of some common aromatic coupling systems.*

A single substituent that is either strongly shielding or deshielding (for example —$COCH_3$, in acetophenone, IV) usually causes the *ortho* protons to move upfield or downfield with respect to the *meta* and *para* protons. We usually observe a two-proton complex multiplet separated from a three-proton complex multiplet (see figure 3.27).

Identical groups *ortho* to each other produce considerable symmetry in the molecule, and the n.m.r. spectrum is usually complex but symmetrical about the mid-point of the multiplet (see figure 3.27). Such spectra are also AA′BB′ systems, but with more complexity than the simplest of the corresponding *para* systems (for example diethyl phthalate, V).

Highly substituted rings may produce very simple spectra, and in a great number of cases it is possible to use the coupling constants to confirm orientation, since $J_{ortho} > J_{meta} > J_{para}$.

Heterocyclic aromatic systems are, for the most part, governed by similar rules to benzenoid compounds. Clear first-order spectra should be analysed completely, and complex multiplets should be treated with caution.

Polynuclear aromatic hydrocarbon systems invariably give very complex spectra, the chemical shift positions ranging from 7–9 δ.

Vinyl and allyl systems. Many vinyl systems give rise to AMX coupling (see figure 3.28) especially where the substituent is an oxygen function as in vinyl acetate. To understand the coupling we should recall that H_A and H_M are *trans* and *cis*, respectively, to the substituent, and this often means that they have widely different chemical shift positions. The *trans* coupling J_{MX} is usually larger than the *cis* coupling J_{AX}; J_{AM}, being a *gem* coupling, is also large. In the absence of a strongly electronegative substituent, vinyl spectra will have ABC complexity and only approximate chemical shift data should be extracted.

Altogether four proton environments couple together in allyl systems, and this produces extremely complex multiplets. A very large, three-proton multiplet may extend almost from 4.5–6.5 δ, generated by the alkene protons.

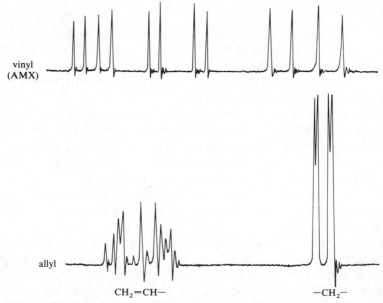

The methylene protons may appear superficially as a doublet because of coupling to H_A, but the small allylic couplings (0–2 Hz) to protons H_B or H_C may also be observed under closer examination. The chemical shift positions for the CH_2 group can be predicted fairly reliably from table 3.6. The typical appearance of an allyl coupling system is shown in figure 3.28.

Figure 3.28 *The appearance of representative vinyl and allyl coupling systems.*

Virtual coupling. The methyl group of a long alkyl chain $(\ldots CH_2CH_2CH_2CH_3)$ should appear as a simple triplet in the n.m.r. spectrum, but more commonly it appears as a very blurred triplet indeed. The broadening of the CH_3 signal is attributed to interaction with the numerous non-neighbour CH_2 groups, even though there is no true coupling with these. True coupling only occurs with the adjacent CH_2 group, but this group serves to link spin interactions with the other methylenes to the terminal methyl. Such a higher-order effect is termed *virtual coupling*, and is responsible for a number of distortions in what one might have expected to be first-order spectra.

For virtual coupling to arise in protons P, they must be strongly coupled to their neighbour protons Q, which must in turn be strongly coupled to their neighbours R; the coupling between P and R must also be zero.

3.11 SIMPLIFICATION OF COMPLEX SPECTRA

If a non-first-order spectrum is obtained at 60 MHz, a number of techniques can be applied to simplify the complexity and enable more accurate analyses of chemical shift and coupling constant data to be made.

3.11.1 INCREASED FIELD STRENGTH

We have seen that first-order multiplicity is usually produced when $\Delta\delta \geqslant 6\,J$. Chemical shift positions (when measured in Hz) are field dependent: the methyl resonance in acetates appears at 2.0 δ, or 2 p.p.m. downfield from TMS. In a 60 MHz instrument (1.4 T) 2 p.p.m. corresponds to 120 Hz, while in a 100 MHz instrument (2.3 T) 2 p.p.m. corresponds to 200 Hz.

Coupling constants are, however, independent of field strength, so that the ratio $\Delta\delta : J$ is effectively increased as the field strength is increased from 1.4 to 2.3 T. If two coupling multiplets overlap at 1.4 T, we can pull the multiplets apart by increasing the magnetic field. The further we pull the multiplets apart, the more likely is the spectrum to approach first-order, since we are in effect increasing $\Delta\delta$ with respect to J.

Figure 3.29 shows the n.m.r. spectrum of 4-chlorobutyric acid recorded at 60 MHz, 100 MHz and 220 MHz. At 60 MHz and 100 MHz the signal from the central CH_2 group is non-first-order, while at 220 MHz the first-order quintet multiplicity is shown.

If a compound gives a complex n.m.r. spectrum at 60 MHz, it will normally be improved by recording at 100 MHz or 220 MHz, although the degree of improvement depends on the particular chemical shift differences and coupling constants involved.

3.11.2 SPIN DECOUPLING OR DOUBLE RESONANCE

Multiplicity of signals arises because neighbouring protons have more than one spin orientation (low energy or parallel, and high energy or anti-parallel): proton A in figure 3.30 appears as a doublet because of the two

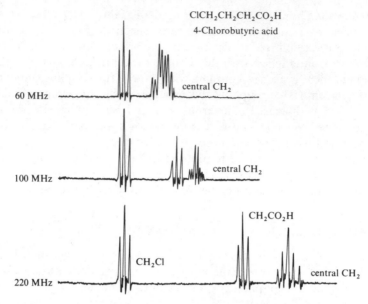

Figure 3.29   *Representation of the n.m.r. spectrum of 4-chlorobutyric acid at 60, 100 and 220 MHz.*

spin orientations of X. If we irradiate X with the correct radiofrequency energy, we can stimulate rapid transitions (both upwards and downwards) between the two spin states of X so that the lifetime of a nucleus in any one spin state is too short to resolve the coupling with A. If proton A 'sees' only one time-averaged view of X, than A will come to resonance only once, and not twice. By the same argument, if we irradiate proton A with the correct radiofrequency energy to cause it to undergo rapid transitions between its two spin states, proton X will only 'see' one time-averaged view of A, and appear only as a singlet.

We have seen earlier that $\Delta t \cdot \Delta v \approx 1/2\pi$: for a pair of coupled protons, the time needed to resolve the two lines of a doublet (Δt) is related to the separation between the lines, that is the coupling constant (Δv). For the above example, we shall be able to resolve the H_A doublet provided each spin state of H_X has a lifetime greater than Δt. Double irradiation shortens this lifetime to less than Δt, and consequently we are unable to resolve the H_A doublet — which appears therefore as a singlet. See also section 3.8.4.

To perform this operation we require, in addition to the basic n.m.r. instrument, a *second* tunable radiofrequency source to irradiate proton X at the necessary frequency (near to its precession frequency), while recording the remainder of the spectrum as before. Since we are making simultaneous use of two radiofrequency sources, the technique is called *double resonance* or *double irradiation*: since the nuclear spins during the process are 'less coupled' than before, we also call it *spin decoupling*.

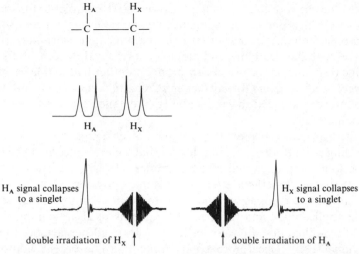

Figure 3.30 *Double irradiation (spin decoupling) of an* AX *spectrum.*

For the method to be successful, the chemical shift positions for the coupling multiplets should be no closer than ≈ 1 p.p.m. Decoupling of non-first-order spectra can frequently lead to first-order spectra, provided this condition is met.

Other applications of the double resonance method (spin tickling and I.N.D.O.R.) are discussed in the supplement.

3.11.3 CONTACT-SHIFT REAGENTS—CHEMICAL SHIFT REAGENTS
The n.m.r. spectrum of 6-methylquinoline ($CDCl_3$ solution) is reproduced in figure 3.31. The lower spectrum is the normal record; the upper spectrum was

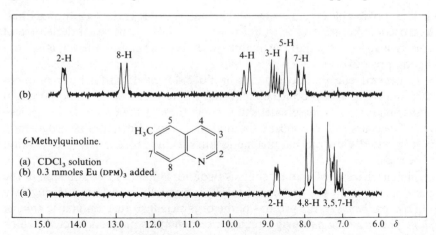

Figure 3.31 *Effect of a contact-shift reagent* ($Eu(DPM)_3$) *on the n.m.r. spectrum of 6-methylquinoline. (Ring protons only)*

recorded after the addition of a soluble europium(III) complex to the solution, and the spectrum is pulled out over a much wider range of frequencies so that it is simplified almost to first-order. The paramagnetic europium(III) ion complexes with the quinoline, and induces enormous downfield shifts in the quinoline resonances. The use of europium and other lanthanide derivatives as *chemical shift reagents* (or *contact shift reagents*) is a recent innovation, but has already extended the applicability of detailed n.m.r. studies to very complex molecules.

For the method to succeed, the organic molecule must be able to donate nonbonding pairs to the europium ion so that we are concerned principally with the following functional classes: amines, alcohols, ketones and aldehydes, ethers and thioethers, esters, nitriles and epoxides.

The lanthanide complex should be soluble in common n.m.r. solvents for wide applicability, and those most frequently used are complexes with two enolic β-diketones, *di*pivaloyl*methane* (DPM) and hepta*f*luorodimethyl-*o*ctane*d*ione (FOD). Deca*f*luorohepatane*d*ione (FHD), a recent addition, is also soluble in CCl_4.

$$
\left[\begin{array}{c}
\quad\quad R_1 \\
O=C \\
\quad\quad CH \\
Eu \\
O-C \\
\quad\quad R_2
\end{array} \right]_3
$$

$$R_1 = R_2 = -CMe_3 \equiv Eu(DPM)_3$$
$$R_1 = -CMe_3, R_2 = -C_3F_7 \equiv Eu(FOD)_3$$
$$R_1 = R_2 = -C_2F_5 \equiv Eu(FHD)_3$$

The FOD complexes are more soluble than the DPM variants, although their hygroscopic nature can be troublesome. The fully deuteriated derivative of FOD is also commercially available, and this eliminates the signals from ligand protons.

In general, europium complexes produce downfield shifts, while praeseodymium complexes produce upfield shifts. Ytterbium, erbium and holmium compounds tend to give greater shifts, but in these last two some line broadening also occurs and complicates the analysis of multiplets: the line broadening is associated with the paramagnetic ion's ability to accelerate relaxation processes.

The mechanism of contact shifts is twofold. Unpaired electron spin in the paramagnetic ion (for example Eu(III)) is partially transferred *through the intervening bonds* to the protons of the organic substrate; this is true *contact shift*. The spinning paramagnetic ion also generates magnetic vectors which operate *through space* and create secondary fields around the protons; this is *pseudocontact shift*, and predominates in the case of the lanthanide ions.

3.11.4 COMPUTER FOR AVERAGING OF TRANSIENTS (C.A.T.)

With low concentrations, for example when working with solutions containing less than 5 per cent of solute, the signal-to-noise ratio on the n.m.r. spectrum becomes lower, since it is necessary to work at higher amplifications. At significantly lower concentrations it may ultimately become difficult to distinguish true n.m.r. peaks from the noisy base-line. Many biological materials are only isolated in small quantities, and n.m.r. studies are severely handicapped by this; the problem also applies to compounds that are sparingly soluble in common n.m.r. solvents.

Movements of the recorder pen caused by noise are random: over a series of scans therefore the sum of all noise signals will be zero. If we can store all the signals coming from the machine in a memory bank, then we can average out all the transient noise signals to zero; meanwhile all *true* n.m.r. signals from the sample will appear at exactly the same place in the spectrum, and the algebraic summing of these signals will lead to signal enhancement.

A small computer attached to the n.m.r. instrument can perform this *computer averaging of transients* (C.A.T.); the equipment performs a digital summation process, and can also be called a *digital signal averager* (D.S.A.).

The use of the C.A.T. can be extended to study nuclei that have low abundance and/or small magnetic moment (for example ^{13}C, ^{14}N, etc.). Summation of a very large number of spectra runs is required before good ^{13}C spectra are obtained from the natural abundance (1.1%), and pulsed n.m.r. renders this application obsolescent (see supplement 3, Fourier transform n.m.r.). The improvement in signal-to-noise ratio is proportional to (the number of scans)$^{1/2}$.

3.12 TABLES OF DATA

In using these tables, τ values can be calculated from $\tau = 10 - \delta$, but additional *shifts* mentioned in relation to δ units are in the opposite direction for τ units (for example in tables 3.5, 3.6 and 3.9, *reverse* the signs for the expected influence on τ units).

Protons of NH, OH *and* SH *groups* (whose δ values are listed in table 3.8) show special characteristics. All are removed by deuteration (see page 115) and all are affected by solvent, temperature and concentration (see page 98). Signals for ROH protons will appear as a singlet or as a multiplet, depending on whether coupling to neighbour protons is observed (see page 109). Primary amines with concentrated sulphuric acid are completely protonated (to RNH_3^+): proton exchange is suppressed, and the signal 'disappears' because of coupling to ^{14}N, with $J_{NH} \approx 50$ Hz (see page 114). Secondary and tertiary amines with concentrated sulphuric acid give a sharpened line at low field, because ^{14}N relaxation is so rapid that no N—H coupling is observable.

Table 3.4 δ values for the protons of CH_3, CH_2 and CH groups attached to groups X, where R = alkyl, and Ar = aryl

X	$CH_3 X$	$R'CH_2 X$	$R'R''CHX$
—R	0.9	1.3	1.5
—CH^b (with CH$_2^a$ / O ring)	1.3	a 3.5	b 3.0
\=	1.7	1.9	2.6
=—=—=, etc. (i.e., end-of-chain)	1.8		
=—=—=, etc. (i.e., in-chain)	2.0	2.2	2.3
=N—	2.0	—	—
—≡	2.0	2.2	—
—COOR, —COOAr	2.0	2.1	2.2
—CN	2.0	2.5	2.7
—CONH$_2$, —CONR$_2$	2.0	2.0	2.1
—COOH	2.1	2.3	2.6
—COR	2.1	2.4	2.5
—SH, —SR	2.1	2.4	2.5
—NH$_2$, —NR$_2$	2.1	2.5	2.9
—I	2.2	3.1	4.2
—CHO	2.2	2.2	2.4
—Ph	2.3	2.6	2.9
—Br	2.6	3.3	4.1
—NHCOR, —NRCOR	2.9	3.3	3.5
—CI	3.0	3.4	4.0
—OR	3.3	3.3	3.8
—NR$_3$	3.3	3.4	3.5
—OH	3.4	3.6	3.8
—OCOR	3.6	4.1	5.0
—OAr	3.7	3.9	4.0
—OCOAr	3.9	4.2	5.1
—NO$_2$	4.3	4.4	4.6

Table 3.5 Influence of functional group X on the chemical shift position (δ) of CH_3, CH_2 and CH protons β to X

X	For β-shifts, add the following to the δ values given in table 3.4		
	CH_3—C—X	CH_2—C—X	CH—C—X
—C=C	0.1	0.1	0.1
—COOH, —COOR	0.2	0.2	0.2
—CN	0.5	0.4	0.4
—CONH$_2$	0.25	0.2	0.2
—CO—, —CHO	0.3	0.2	0.2
—SH, —SR	0.45	0.3	0.2
—NH$_2$, —NHR. —NR$_2$	0.1	0.1	0.1
—I	1.0	0.5	0.4
—Ph	0.35	0.3	0.3
—Br	0.8	0.6	0.25
—NHCOR	0.1	0.1	0.1
—Cl	0.6	0.4	0
—OH, —OR	0.3	0.2	0.2
—OCOR	0.4	0.3	0.3
—OPh	0.4	0.35	0.3
—F	0.2	0.4	0.1
—NO$_2$	0.6	0.8	0.8

Table 3.6 δ values for the protons of CH_2 (and CH) groups bearing more than one functional substituent (modified Shoolery rules)

Note:

for H_2C $\overset{X^1}{\underset{X^2}{<}}$ $\delta CH_2 = 1.2 + \Sigma a$

Less accurate for $HC \overset{X^1}{\underset{X^3}{—X^2}}$

X	a	X	a
—=	0.75	—Ph	1.3
—≡	0.9	—Br	1.9
—COOH, —COOR	0.7	—Cl	2.0
—CN, —COR	1.2	—OR, —OH	1.7
—SR	1.0	—OCOR	2.7
—NH$_2$, —NR$_2$	1.0	—OPh	2.3
—I	1.4		

Table 3.7 δ values for H attached to unsaturated and aromatic groups

Structure	δ	Structure	δ
H—≡—R	1.8*	(cyclohexene ring with H)	5.6
H—≡⌐OH	2.4*	(H on C=C–C=O)	5.8
H—≡—=—, etc.	2.7*	(H on C=C–C=O)	6.0
H—≡—Ph	2.9*	(H, H on C=C–C=O)	6.2
H—≡—CO—	3.2*	Ph—=⟨H / CO— (cis or trans)	6.6
H₂C=C⟨R / R′	4.6	H / Ph =—CO—	7.8
H₂C=—=—, etc.	4.9	⟩N—C⟨H / =O	7.8
H / R =—OR′ (acyclic)	5.0		
H / R (=CMe₂)	5.3		
H / Ph =⟨H	5.0		
Ph / H (=)	5.3		
R—=—=— (in-chain), H	6.2		

Table 3.8 δ values for the protons of OH, NH and SH groups*

ROH	0.5–4.0	Higher for enols (11.0–16.0)
		Lines often broadened
ArOH	4.5	Raised by hydrogen bonding to ≈9.0
		Chelated OH, ≈11.0
RCOOH	10.0–13.0	
RNH₂, RNHR′	5.0–8.0	Lines usually broadened
ArNH₂, ArNHR′	3.5–6.0	Occasionally raised. Lines usually broadened
RCONH₂, RCONHR′	5.0–8.5	Lines frequently very broad, and even unobservable
RCONHCOR′	9.0–12.0	Lines broadened
RSH	1.0–2.0	
ArSH	3.0–4.0	
=NOH	10.0–12.0	Often broadened

* See notes on page 125.

Table 3.7 *continued*

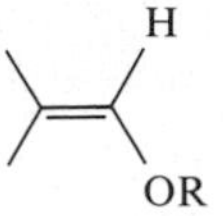

H
| | 6.8
OR

ROC(=O)H   8.0

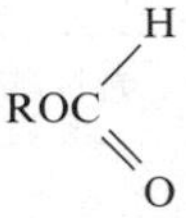

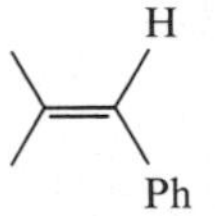

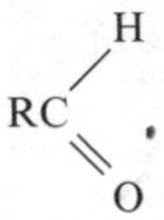

H
| | 7.0
Ph

RC(=O)H   9.6

Ph—H   7.27
(see table 3.9)

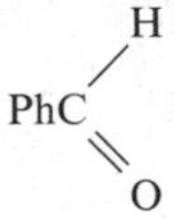

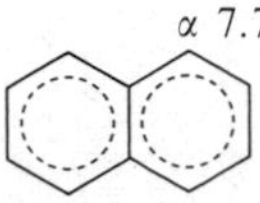

PhC(=O)H   9.9

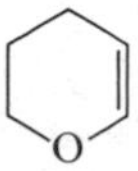

α 7.7
β 7.5 (naphthalene)

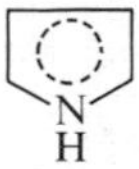

γ 7.4
β 7.0
α 8.5 (pyridine)

4.5
6.2 (dihydropyran)

β 6.1
α 6.5 (pyrrole, N—H)

β 7.1
α 7.2 (thiophene)

β 6.3
α 7.4 (furan)

* Alkyne proton signals removed on deuteration, and δ values increased by a trace of pyridine.

Table 3.9 Shifts in the position of benzene protons (7.27 δ) caused by substituents

Substituent	ortho	meta	para
—CH$_3$	−0.15	−0.1	−0.1
—=	+0.2	+0.2	+0.2
—COOH, —COOR	+0.8	+0.15	+0.2
—CN	+0.3	+0.3	+0.3
—CONH$_2$	+0.5	+0.2	+0.2
—COR	+0.6	+0.3	+0.3
—SR	+0.1	−0.1	−0.2
—NH$_2$	−0.8	−0.15	−0.4
—N(CH$_3$)$_2$	−0.5	−0.2	−0.5
—I	+0.3	−0.2	−0.1
—CHO	+0.7	+0.2	+0.4
—Br	0	0	0
—NHCOR	+0.4	−0.2	−0.3
—Cl	0	0	0
—NH$_3$$^+$	+0.4	+0.2	+0.2
—OR	−0.2	−0.2	−0.2
—OH	−0.4	−0.4	−0.4
—OCOR	+0.2	−0.1	−0.2
—NO$_2$	+1.0	+0.3	+0.4
—SO$_3$H, —SO$_2$Cl, —SO$_2$NH$_2$, etc.	+0.4	+0.1	+0.1

Table 3.10   Proton–proton spin-coupling constants

Function	J_{ab}/Hz
(gem)	10–18 depending on the electronegativities of the attached groups
CH_a—CH_b (vic)	depends on dihedral angle: see section 3.9.3
C=C H_a, H_b	3–7
C=C (cis)	5–14
C=C (trans)	11–19
C=C, C—H_a	4–10
H_a—C=C, C—H_b (cis or trans)	0–2 (for aromatic systems, 0–1)
C=CH_a—CH_b=C	10–13
H_a —H_b	ortho, 7–10 meta, 2–3 para, 0–1

Table 3.11 Solvents for n.m.r. work

Solvent	Approximate δ for ^1H equivalent (as contaminant)	b.p./°C	f.p./°C
acetic acid-d_4	13 and 2	118	16.6
acetone-d_6	2	56	−95
acetonitrile-d_3	2	82	−44
benzene-d_6	7.3	80	5.5
carbon disulphide	—	46	−108.5
carbon tetrachloride	—	77	−23
chloroform-d	7.3	61	−63
deuterium oxide	5	101.5	3.8
dimethylsulphoxide-d_6	2	189	18
methanol-d (CH_3OD)	3.4	65	−98
hexachloroacetone	—	203	−2
pyridine-d_5	7.5	115	−42
toluene-d_8	7.3 and 2.4	110	−95
trifluoroacetic acid-d	13	72	−15

SUPPLEMENT 3

3S.1 DOUBLE RESONANCE TECHNIQUES

Straightforward spin decoupling (see section 3.11.2) involves irradiating a group of protons with sufficiently intense radiofrequency energy to eliminate almost completely the observed coupling to neighbour protons. By using lower-intensity radiofrequency signals in the double-resonance experiment, we can effect perturbations in the spin populations or energy levels without causing complete decoupling.

A full description in classical terms of the observed changes during double irradiation is impossible, but we can say that weak double irradiation in one multiplet may cause (i) a simplification of a related multiplet or (ii) the appearance of *further splittings* in the lines of the related multiplets by the introduction of additional possible energy transitions or (iii) changes in the line intensities of related multiplets through disturbances in the spin populations.

One important result of these double resonance techniques is the ability to determine the signs of coupling constants (see section 3.8.3); in suitable cases the sign of *J* can be related to structure, since vicinal couplings are generally positive, while geminal and allylic couplings are usually negative. In practice it is easier to determine the *relative* signs of *J* values than it is to measure the absolute sign, and the examples below show two such determinations (spin tickling, I.N.D.O.R.).

A second important result of these double resonance techniques is the identification of coupled protons in spectra that are too complex for detailed analysis. This application is likely to be of great assistance to organic chemists working with complex molecules (I.N.D.O.R.).

3S.1.1   Spin tickling

Figure 3.32 represents schematically a typical twelve-line AMX spectrum, and show the effect of *strong* decoupling irradiation near the origin chemical shift position for proton A; coupling of A with M and X is eliminated, so that only the small J_{MX} appears in the signals of protons M and X. For the sake of clarity, $J_{AM} \gg J_{AX} \gg J_{MX}$.

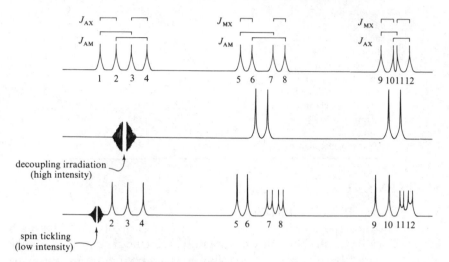

Figure 3.32   *Effects of decoupling and spin tickling on an* AMX *spectrum.*

Spin tickling is carried out by *weakly* irradiating *one line* of the A spectrum, and observing that, for example, in this case, lines 7 and 8 of the M signal and lines 11 and 12 of the X signal are further split (*not* decoupled). The intensity of the double irradiation in spin tickling is typically about one quarter that needed for 'complete' decoupling.

Now lines 5 and 6 are associated with one particular spin orientation of A, while lines 7 and 8 are associated with the opposite spin orientation of A; conversely lines 5 and 7 are associated with one spin orientation of X, while lines 6 and 8 are associated with the opposite spin orientation of X.

In the X part of the spectrum, lines 9 and 10 arise from one spin orientation of A, and lines 11 and 12 arise from the opposite spin orientation of A; finally, lines 9 and 11 are associated with one spin orientation of M, while lines 10 and 12 are associated with the opposite orientation of M.

We have to decide which orientations are related: does the (+) orientation of A give rise to the set of lines 5, 6 and 9, 10 (in which case the (−) orientation gives rise to 7, 8 and 11, 12)? The other alternative is that the (+) orientation of A gives rise to the set of lines 5, 6 and 11, 12 (in which case the (−) orientation gives rise to 7, 8 and 9, 10).

Spin tickling on line 1 has been shown to perturb lines 7, 8 and 11, 12, so that these lines must all arise from the *same* spin orientation of A; the signs of J_{AM} and J_{AX} must be the same, and opposite to that of J_{MX}.

3S.1.2   Internuclear double resonance (I.N.D.O.R.)

Modern instruments of high stability have permitted the development of this particular double-resonance technique to a stage where it can be carried out in a semi-routine way; the method gives the same information as spin tickling and, because of its greater selectivity and sensitivity, will probably displace this.

It was originally used for studying nuclei such as ^{13}C, which has low natural abundance and small magnetic moment. Scanning through the ^{13}C resonance frequency with a low-intensity decoupling signal generates changes in the associated ^1H spectra. Much information could be deduced without directly observing the ^{13}C resonance, and this gave rise to the name *internuclear double resonance*, or I.N.D.O.R.

A similar technique is used for proton–proton I.N.D.O.R. experiments; namely one signal is observed while other coupling nuclei are double irradiated.

In the I.N.D.O.R. experiment applied to simple AX systems, we monitor the *line intensities* of the A signal and sweep a weak radiofrequency source through the X frequencies; as the X frequencies are stimulated, line intensities in the A signal will alter, and we can plot this change in intensity onto a recorder. The perturbing irradiation used in I.N.D.O.R. is about one twentieth that required for complete decoupling, and about one fifth that for spin tickling. The perturbed signals and the monitored signals can therefore be very much closer on the spectrum than is true in decoupling or spin tickling. In I.N.D.O.R., only the *populations* of spin states are altered—no change in energy *levels* is induced.

A simple I.N.D.O.R. experiment is illustrated in figure 3.33 for a four-line AX spectrum. We continuously monitor the line intensity (peak height) of line A_1, and sweep the perturbing irradiation through X_1 and X_2; at X_1, the perturbation causes a *decrease* in the line intensity of A_1, while at X_2 it causes an *increase*. If we monitor line A_2 and repeat the irradiation of X lines, we get a second I.N.D.O.R. spectrum showing that the line intensity of A_2 *increases* for irradiation at X_1 and *decreases* for irradiation at X_2.

The energy diagram in figure 3.34 (compare figure 3.16) helps to explain why line intensities change in the A signal during double irradiation of the X signal. In this diagram, we represent each line on the spectrum as a transition between spin states 1, 2, 3 and 4 (noting that energetically $A_1 > A_2 > X_1 > X_2$).

Irradiation at X_1 will only produce an I.N.D.O.R. signal from line A_1 if they have an energy level in common; from the diagram we see this to be so (level 1). Irradiation at X_1 induces the transition $3 \rightarrow 1$, and the population

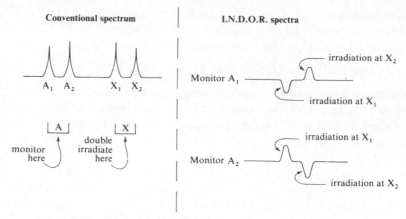

Figure 3.33   *Typical I.N.D.O.R. spectra for an AX coupling system.*

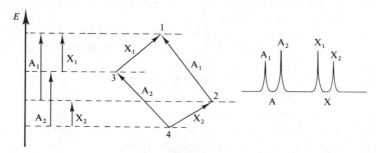

Figure 3.34   *Energy diagram for I.N.D.O.R. study of a typical AX coupling system.*

of spin state 1 is increased. Because of this, the probability of transition $2 \to 1$ (the A_1 line) becomes less favoured, and the line intensity of A_1 is reduced (negative I.N.D.O.R. signal).

Conversely, because the transition $3 \to 1$ is stimulated, the population of spin state 3 is reduced, and therefore transition $4 \to 3$ (the A_2 line) becomes more probable; the A_2 line intensity increases (positive I.N.D.O.R. signal).

We can follow this by considering irradiation at line X_2, which stimulates transition $4 \to 2$; the populations of spin states 4 and 2 are perturbed, state 4 being depleted and state 2 augmented. The transitions that share these energy levels are also affected: the A_2 line intensity will decrease (negative I.N.D.O.R. peak) and the A_1 line intensity will increase (positive I.N.D.O.R. peak).

We describe the relationship between the transition X_2 and A_1, and between A_2 and X_1 as being *progressive*; the relationship between transitions A_2 and X_2, and between X_1 and A_1 are called *regressive*. *Positive* I.N.D.O.R. signals are produced when monitored and irradiated lines have a *progressive* relationship: *negative* I.N.D.O.R. peaks arise when monitored and irradiated lines have a *regressive* relationship.

By ingenious instrumentation it can be arranged that the I.N.D.O.R. spectra are recorded directly onto the same trace as the normal spectrum in such a way that the I.N.D.O.R. peaks line up with the original transitions that are being affected. This factor is most important in complex spectra, an example of which is reproduced in figure 3.35.

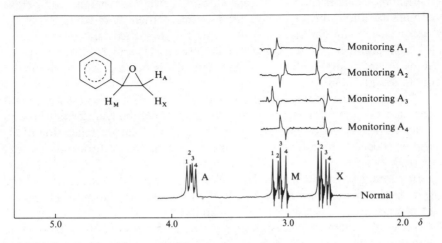

Figure 3.35 *Normal and I.N.D.O.R. n.m.r. spectra of styrene oxide (epoxide ring protons only). (Recorded at 90 MHz.)*

Identification of coupling protons. If one line of a complex spectrum is monitored, and an unresolvable multiplet swept by the perturbing radiation, I.N.D.O.R. peaks will be recorded at those points within the complex multiplet where there are signals coupled to the monitored line.

In most cases the coupling constants can also be deduced, since the spacing between I.N.D.O.R. signals will be the same as J (unless other couplings are also present). Because the irradiating intensity is so low in I.N.D.O.R., much less disturbance to adjacent signals is produced than in other double resonance techniques, so that more precise and selective information about closely spaced multiplets can be obtained.

Signs of the coupling constants. In figure 3.36 (styrene oxide), monitoring of line A_1 shows I.N.D.O.R. signals from lines M_1, M_3, X_1 and X_2, which means that M_1 and M_3 correspond to the same spin orientation of X, and that X_1 and X_2 correspond to the same spin orientation of M. (The I.N.D.O.R. method is being used to detect those lines having common energy levels.) Since, therefore, lines A_1 and M_1 have the same spin orientation of X in common, J_{AX} and J_{MX} have the same sign. By the same argument, lines A_1 and X_1 correspond to the same spin orientation of M, so that J_{AM} and J_{MX} have the same sign.

3S.1.3   Nuclear Overhauser effect (N.O.E.)

The nuclear Overhauser effect (N.O.E.) can be used to demonstrate that two protons (or groups of protons) are in close proximity within the molecule, and is therefore of considerable value in the study of molecular geometry.

The basic observation of N.O.E. can be described by reference to the hypothetical molecule I. The two protons H_a and H_b we must imagine to be close enough to allow through-space interaction of their fluctuating magnetic vectors; each can contribute to the other's spin–lattice relaxation process (T_1). The number of intervening bonds between H_a and H_b is too large to allow normal coupling between them, but this is not a prerequisite for the operation of N.O.E.

If we double irradiate at the H_b signal, we shall stimulate absorption and emission processes for H_b, and this stimulation will be transferred through space to the relaxation mechanism of H_a. The spin–lattice relaxation of H_a will be speeded up, leading to a net increase in the n.m.r. absorption signal of H_a.

In summary, provided that H_b makes a significant contribution to the spin–lattice relaxation process of H_a, then double irradiation of H_b causes an increase in the intensity of the H_a signal (by 15–50%).

Since molecular oxygen is paramagnetic, and can contribute to nuclear spin–lattice relaxation processes, the N.O.E. experiment should be carried out on deoxygenated samples to avoid this possible complication.

A simple example of N.O.E. is found in isovanillin. If we first record the n.m.r. spectrum for isovanillin normally, and then while irradiating at the CH_3O frequency, the integral for the *ortho* proton (which appears as a doublet) is markedly increased.

I   isovanillin

A practical consequence of N.O.E. is that in spin decoupling experiments the line intensities observed on a decoupled spectrum may not be the same as in the normal (nondecoupled) spectrum, and these intensities may not always correspond to integral numbers of protons. This same observation applies in the proton-decoupled spectra of ^{13}C resonances (see page 142).

3S.2   VARIABLE-TEMPERATURE N.M.R.

Valuable structural information can be obtained by recording the n.m.r. spectra of organic compounds at high and low temperatures, when rotational and intermolecular forces can be grossly changed. The practical difficulties in introducing a heated or cooled sample into the n.m.r. magnet are considerable, since magnetic field changes with magnet temperature, but insulated probes have now been developed that allow variable-temperature n.m.r. to be recorded routinely.

3S.2.1 The variable-temperature probe

Two main systems are in use to produce high- or low-temperature environments for the sample in the n.m.r. spectrometer. In both systems, high temperatures are achieved by passing a gas (for example air or nitrogen) through a heater (with suitable thermostatting) and then around the sample tube. Low temperatures are produced in one design by evaporation of a controlled quantity of coolant (for example liquid nitrogen); the other design utilises the Joule–Thomson effect, the gas being passed directly from the storage cylinder through a small orifice within the probe.

3S.2.2 Applications

A number of observations throughout this chapter have indicated the practical value of variable-temperature n.m.r. (see pages 96, 98).

It is not difficult to see its application to studies in restricted rotation or ring inversions, etc., and there is a widely recognised advantage in the ability to record the n.m.r. spectra of thermolabile materials, which are too transient for normal-temperature studies. Many materials (for example polyethylene, isotactice polypropylene) are much more easily studied at high temperature,

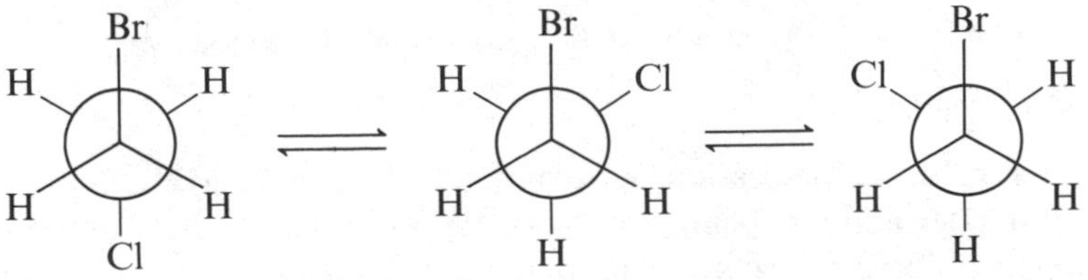

halogenoethanes

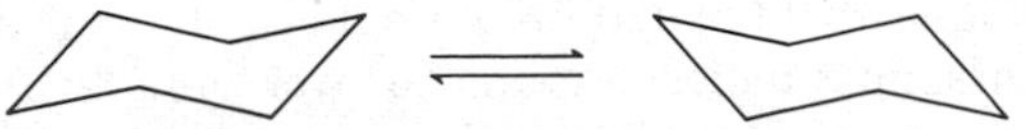

cyclohexanes

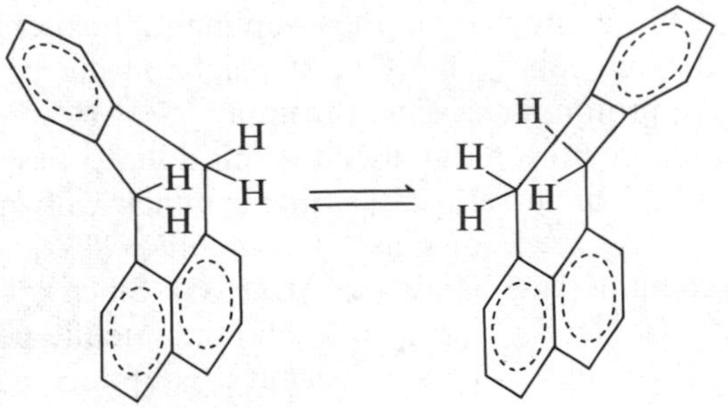

pleiadenes

and may give relatively sharp spectra because of the attendant decrease in viscosity.

A few of the molecules whose rotational and inversional processes have been studied are shown on page 137. By noting the temperature at which such interconversions begin to occur (by gradually heating a sample in which these processes have been 'frozen') it is possible to calculate the energies of activation for the rotations, etc.

Note that the protons of a methylene group adjacent to a chiral centre will never show equivalence by increasing the temperature of the sample.

H_a and H_b are diastereotopic

No matter their position, H_a and H_b will always be in different chemical environments, and are termed *diastereotopic*. Chiral molecules (enantiomers) normally give rise to the same spectra, even though their absolute stereo-chemistries are different, but these absolute differences may be studied by using chiral solvents or chiral lanthanide-shift reagents (for example derived from camphor) to induce different magnetic environments into the otherwise mirror image forms.

3S.3 ^{13}C N.M.R.—Pulsed N.M.R. and Fourier Transforms

3S.3.1   Introduction

In a 2.3 T field, the precession frequency of ^{13}C is ≈ 25 MHz, that for ^{1}H being 100 MHz and ^{12}C being nonmagnetic. In principle, therefore, it is not difficult to observe ^{13}C n.m.r. The magnetic moment of ^{13}C is about one quarter that of ^{1}H, so that signals are inherently weaker, but the over-whelming problem is that the natural abundance of ^{13}C is only 1.1 per cent. The problem in simple molecules can be overcome by synthesising ^{13}C-enriched samples, but this is of little value in complex molecules.

Historically, ^{13}C was studied via the coupling that exists between the 1.1 per cent natural abundance of ^{13}C and neighbouring protons: ^{13}C has $I = \frac{1}{2}$, so that the proton n.m.r. signal from a ^{13}C—^{1}H bond is split into a doublet. Many intense proton signals show this doublet (less than 1 per cent of the intensity of the main ^{1}H peak) sitting symmetrically astride the main peak: these tiny peaks are known as ^{13}C *satellite peaks*, and are not to be confused with spinning side-bands (see page 85). Since ^{13}C couplings are large ($J_{^{13}C-^1H} \approx 100$ Hz, $J_{^{13}C-C-^1H} \approx 25$ Hz, the satellite peaks are usually far enough away from the ^{1}H peak to permit decoupling and I.N.D.O.R.

One solution to the low natural abundance and small magnetic moment of ^{13}C is to use the C.A.T. (see page 125) to scan a large number of successive runs, but even a tenfold signal-to-noise enhancement demands one hundred stored spectra, each requiring perhaps three minutes to scan. Clearly time is against this technique.

3S.3.2   Pulse techniques

In the normal mode of instrument operation, spectra require several minutes to record because each transition is induced in succession by continuous scan from low field to high field, the radiofrequency signal remaining constant: this is the *continuous wave* mode of operation (C.W.).

If we could stimulate all transitions *simultaneously*, the same transitions within the nuclear spin states would arise, but the excitation would require a mere fraction of a second to execute. This is precisely what is done in *pulsed* n.m.r.: for proton n.m.r., the sample is irradiated (at fixed field) with a strong pulse of radiofrequency energy containing all the frequencies over the ^1H range (for example spread around 100 MHz at 2.3 T, etc.). The protons in each environment absorb their appropriate frequencies from the pulse, and couple with one another in the normal way to create all the coupling sublevels of energy.

The pulse duration may be ≈ 30 µs, and when it is switched off the nuclei undergo relaxation processes and re-emit the absorbed energies and coupling energies. The instrumental problem is that all these re-emitted energies are emitted simultaneously as a complex interacting pattern, which decays rapidly. Figure 3.36 shows a typical pattern of re-emitted decay signals: this output is digitalised in a computer, and each *individual* frequency is filtered out from the complex pattern in much the same way as audio-frequency signals are filtered out from a modulated radio wave in a radio receiver. Each of these frequencies is plotted on a linear frequency scale, which corresponds to the conventional n.m.r. spectrum: this is shown schematically in figure 3.36.

3S.3.3   Fourier transforms

The presentation in (a), figure 3.36, is time-based, and is described by physicists as being in the *time domain*. The transformation to (b), which is frequency-based and is therefore in the *frequency domain*, can be carried out mathematically by *Fourier transforms*. For this reason, *pulsed n.m.r.* is usually called *Fourier transform n.m.r.* (F.T.), to distinguish it from the more conventional continuous wave (C.W.) operation.

3S.3.4   Advantages of F.T. n.m.r.

An entire spectrum can be recorded, computerised and transformed in a few seconds; with a repetition every two seconds for example, 400 spectra can be accumulated in ≈ 13 minutes, giving $20 \times$ signal enhancement.

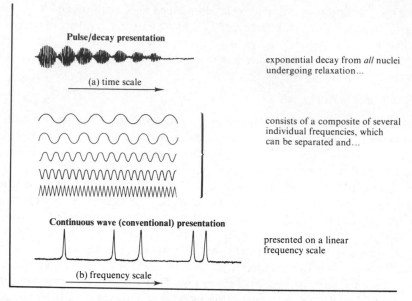

Pulse/decay presentation

exponential decay from *all* nuclei undergoing relaxation...

(a) time scale

consists of a composite of several individual frequencies, which can be separated and...

Continuous wave (conventional) presentation

presented on a linear frequency scale

(b) frequency scale

Figure 3.36   *Schematic representation of pulsed n.m.r.: the output in the time domain is converted to the frequency domain by Fourier transform.*

Technically, pulse durations of 1 μs are possible, but after excitation by one pulse, the nuclei cannot be re-excited until T_1 (the spin–lattice relaxation time) has elapsed: repetition times up to 60 s may be required for nuclei with long T_1 values.

Three low-sensitivity problems in n.m.r. are now easily overcome: (a) samples at very low concentrations; (b) n.m.r. studies on nuclei with low natural abundance (for example ^{13}C); (c) n.m.r. studies on nuclei with small magnetic moments (for example ^{13}C).

The first advantage (a), being able to work at very low concentrations in ^1H n.m.r., is important in biological chemistry where only microgram quantities of material may be available. The F.T. method also gives improved spectra (compared with the use of the C.A.T.) for sparingly soluble compounds.

The most dramatic advantage to organic chemists, (b) and (c), is the ability to record excellent spectra from 'ordinary' molecules containing only the natural abundance of ^{13}C nuclei.

An example is shown in figure 3.37 of the ^{13}C spectrum of menthol, and for comparison the ^1H spectrum of menthol is also shown.

Resolution. Each of the ten lines on the ^{13}C spectrum in figure 3.37(b) represents one carbon atom of menthol, and two immediate differences from the ^1H spectrum are apparent: the ^{13}C spectrum is much simpler, and much more highly resolved.

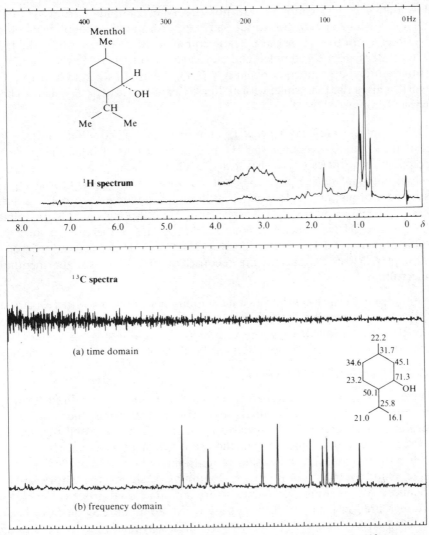

Figure 3.37 *Comparison of the ^{1}H and ^{13}C n.m.r. spectra of menthol (^{13}C chemical shifts in p.p.m. from* TMS*).*

The chemical shift range in the ^{1}H spectrum is only ≈ 4 p.p.m. (240 Hz in this 60 MHz spectrum), while the range in the ^{13}C spectrum is ≈ 80 p.p.m. (2 kHz in this 25 MHz spectrum). Expressed otherwise, the chemical shift differences in the ^{13}C spectrum are about 20 times those shown in the ^{1}H spectrum, and this is typical in all other molecules.

Multiplicity. Both ^{13}C and ^{1}H have $I = \frac{1}{2}$, so that we should expect to see coupling in the spectrum between (a) ^{13}C–^{13}C and between (b) ^{13}C–^{1}H. The probability of two ^{13}C atoms being together in the same molecule is so low that $^{13}C-^{13}C$ couplings are not observed. Couplings from $^{13}C-^{1}H$ interaction have already been discussed (page 138) and these couplings should be observed in the ^{13}C spectra. These couplings would, however, make the ^{13}C spectra extremely complex, and they have been eliminated by decoupling. (Interestingly, if hexadeuteriobenzene, C_6D_6, is used as a standard in ^{13}C work, it gives rise to a triplet with all lines of equal intensity: why should this triplet be observed?)

^{1}H *decoupling—noise decoupling.* To eliminate $^{13}C-^{1}H$ couplings in the ^{13}C spectra, we must decouple the ^{1}H nuclei by double irradiation at *their* resonant frequencies (≈ 100 MHz at 2.3 T). This is an example of *heteronuclear decoupling* (see page 133), but we do not wish merely to decouple specific protons; rather do we wish to double irradiate *all* protons simultaneously while recording the ^{13}C spectrum. A decoupling signal is used that has all the ^{1}H frequencies around 100 MHz, and is therefore a form of radiofrequency noise; spectra derived thus are ^{1}H-*decoupled*, or *noise-decoupled*. Most ^{13}C spectra are recorded in this way, (see the menthol spectrum in figure 3.37).

N.O.E. *signal enhancement.* Since decoupling can interfere with relaxation times, the nuclear Overhauser effect (see page 136) may operate and lead to signal enhancement of certain ^{13}C peaks. The line intensities in the ^{13}C spectrum of menthol are not all equal because of these relaxation effects.

3S.3.5   Structural applications of ^{13}C n.m.r.

Differentiation among alternative organic structures has a long history in ^{1}H n.m.r., and it is substantially extended by ^{13}C n.m.r. Increased shift resolution (compared to ^{1}H spectra) is often sufficient in itself to lead to correct structural assignment, but the use of correlation data for chemical shift positions and the calculation of multiplicity in nondecoupled spectra have their contribution to make. Figure 3.38 shows the approximate chemical shift positions for common organic functional groups: the shifts are measured in p.p.m. from TMS as standard, although CS_2 has also been widely used as a standard. There is a reasonable agreement between chemical shift and hybridisation state in C—C, C—O and C—N bonds, in that sp^2 carbons appear at lowest field (left-hand side of the chart), followed by sp carbons, with sp^3 carbons at highest field.

A simple example of the structural application concerns the dimer of cyclo-octatetraene, which can have either of the structures I, II or III.

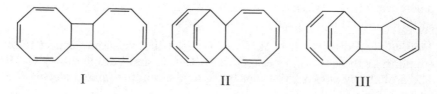

I   II   III

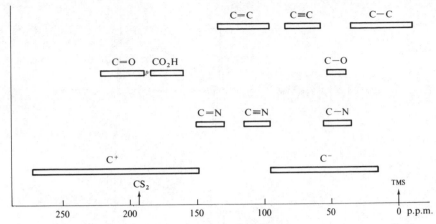

Figure 3.38 *Chemical shift ranges (p.p.m.) for ^{13}C in common organic molecules.*

The ^{13}C spectrum of this dimer showed only four distinct environments of roughly equal intensity, and therefore structure I is correct.

3S.4   CHEMICALLY INDUCED DYNAMIC NUCLEAR POLARISATION
      (C.I.D.N.P.)

When dipropionyl peroxide is thermolysed in a n.m.r. tube in the presence of thiophenol, ethyl radicals are produced, which react with the thiophenol to form thiophenetole

$$CH_3CH_2CO{-}O{-}O{-}COCH_2CH_3 \xrightarrow{\Delta} 2CH_3CH_2CO{-}O\cdot \longrightarrow$$
      dipropionyl peroxide

$$2CH_3CH_2\cdot + 2CO_2$$

$$PhSH + CH_3CH_2\cdot \longrightarrow \longrightarrow \longrightarrow PhSCH_2CH_3$$
      thiophenol   thiophenetole

If the n.m.r. spectrum of the thiophenetole is recorded *during* the course of the reaction, the ethyl signals appear as sketched in figure 3.39(a); the n.m.r. spectrum of normal thiophenetole is also shown at (b) for comparison.

Some of the lines in spectrum (a) have increased intensity (stimulated absorption (A)), while other lines show as emissions (stimulated emission (E)).

We say that the nuclear spins in the product of this radical reaction (that is, the nuclear spins of thiophenetole) are undergoing dynamic polarisation, because of the *chemical reaction* that is producing the molecule. This is an example of *Chemically Induced Dynamic Nuclear Polarisation* (C.I.D.N.P.).

Observation of C.I.D.N.P. effects during a reaction is good evidence that, at least in part, a radical mechanism is involved, and the technique is now

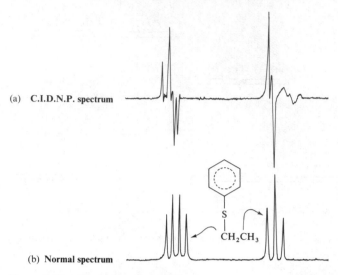

(a)   C.I.D.N.P. spectrum

(b) **Normal spectrum**

Figure 3.39   *Representations of the n.m.r. spectrum of the ethyl group in thiophenetole:* (a) *C.I.D.N.P. spectrum, recorded during the formation of thiophenetole from thiophenol and ethyl radicals,* (b) *normal spectrum.*

being developed rapidly to study formation and decomposition of diaryl-methanes, triarylmethanes, acyl peroxides, etc.

The mechanism of C.I.D.N.P. is complex, and the observed effect (E or A) depends on the way in which particular nuclear spin states interact with the electron spins of the free radicals involved. Additionally, the radicals from a precursor molecule can follow several reaction paths depending on whether the originally formed radical pair react together immediately, or diffuse out from each other to perform other dimerisation or transfer reactions. For a particular radical pair, the singlet and triplet electronic states undergo mixing processes under the influence of nearby nuclear spins; if the resultant mixed electronic state is predominantly singlet in character, the associated nuclear spin states are enhanced in the product formed from the radical pair, and the n.m.r. of the product shows increased absorption for these spin states. If the mixed electronic state is predominantly triplet, emission will be shown for the corresponding nuclear spin states.

The ability to predict or explain whether (E) or (A) will arise in C.I.D.N.P. experiments clearly follows upon the ability to calculate the importance of singlet and triplet contributions to the new energy levels: this skill will not frequently be required of the organic chemist.

3S.5 ^{19}F AND ^{31}P N.M.R.
The concepts of ^{19}F and ^{31}P n.m.r. spectroscopy have been built up contin-uously throughout the main chapter, and the following have been discussed

largely in relation to ^{1}H n.m.r. spectroscopy: precession frequencies and field strengths (page 80); coupling of ^{19}F with ^{1}H (page 113). Proton–fluorine coupling was also demonstrated in the ^{1}H n.m.r. spectrum of 2,2,2-tri-fluoroethanol (figure 3.24).

3S.5.1 ^{19}F n.m.r.

^{19}F is the naturally occurring isotope of fluorine (100% abundance). Apart from the provision of an appropriate radiofrequency source (56.46 MHz at 1.4 T) no major instrument modification is needed to change an n.m.r. spectrometer from ^{1}H work to ^{19}F work.

Chemical shift positions are most often measured from CFCl$_3$ or tri-fluoroacetic acid, and figure 3.40 shows a few of these positions for the most important organic situations. The range of chemical shift covered by even

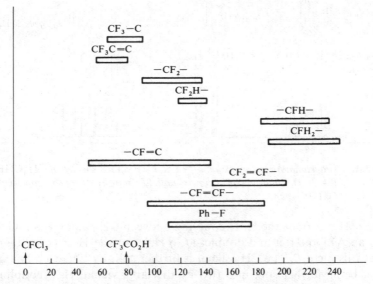

Figure 3.40 *Chemical shift ranges (p.p.m.) for ^{19}F in common organic environments.*

this limited number of fluorine environments is 200 p.p.m., compared with ≈ 10 p.p.m. for the most common ^{1}H positions. In consequence, fluorine resonances tend to be well separated on the spectrum, and first-order spectra are the norm rather than the exception.

Coupling constants between fluorine nuclei cover a wider range than in ^{1}H n.m.r. Geminal F–F coupling ranges from 43–370 Hz, and vicinal F–F from 0–39 Hz; *cis*-fluorines show *J*, 0–58 Hz, and *trans* fluorines show *J*, 106–148 Hz. Long-range coupling over five bonds (through F—C—C—C—C—F) is 0–18 Hz.

Coupling between ^{1}H and ^{19}F is also strong. Geminal coupling ranges from 42–80 Hz, and vicinal ^{1}H–F from 1.2–29 Hz; *cis* ^{1}H–F shows J, 0–22 Hz, and *trans* ^{1}H–F shows J, 11–52 Hz.

Fluorine attached to benzene also couples with protons on the ring, the $J_{H,F}$ ranges being: *ortho*, 7.4–11.8 Hz; *meta*, 4.3–8.0 Hz; *para*, 0.2–2.7 Hz.

Figure 3.41 is a diagrammatic presentation showing the ^{1}H and ^{19}F n.m.r. spectra of 1-bromo-1-fluoroethane (CH_3CHFBr). Each spectrum is recorded using a 1.4 T magnet, so that the ^{1}H spectrum is at 60 MHz, and the ^{19}F spectrum is at 56.46 MHz.

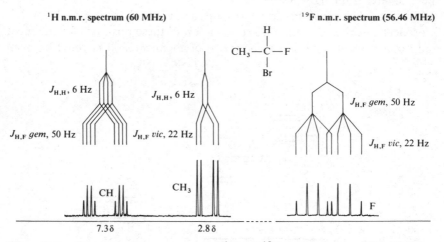

Figure 3.41 *Diagrammatic presentation of ^{1}H and ^{19}F n.m.r. spectra of CH_3CHFBr at 1.4 T. The fluorine spectrum is 'upfield' from hydrogen by approximately 60 000 p.p.m.*

In the ^{1}H spectrum, the H–H coupling between CH_3 and CH should give rise to an AX_3 quartet and doublet (J, 6 Hz), but the H–F coupling complicates this picture. The CH_3 signal is further split by ^{19}F, each line of the doublet being further split into two by the large vicinal H–F coupling (J, 22 Hz); the CH_3 signal is therefore a double doublet. The CH signal is split into a quartet by CH_3 (J, 6 Hz) and then each line is split into two by the enormous geminal H–F coupling (J, 50 Hz); the CH signal is therefore a doublet of quartets.

In the ^{19}F spectrum, the fluorine signal is split into a doublet by the geminal methine proton (J, 50 Hz), and each line is further split into four by the vicinal methyl protons (J, 22 Hz); the fluorine resonance is therefore an overlapping doublet of quartets.

3S.5.2 ^{31}P n.m.r.

Like ^{1}H and ^{19}F, ^{31}P has $I = \frac{1}{2}$ and the multiplicity that it engenders in ^{1}H and ^{19}F spectra is easily predicted using the same rules that we have seen applied to protons and fluorine. We shall restrict ourselves to studying the multiplicity in the ^{1}H and ^{19}F spectra of Me_2PCF_2Me; coupling of phosphorus with Me_2 gives rise to a doublet, and the single Me group is split by ^{31}P into a doublet, and then by the two ^{19}F nuclei so that the Me signal consists of overlapping triplets (see figure 3.42). The fluorine resonance is split into two by the phosphorus, each line being further split into four by the CH_3 protons. If we could observe the ^{31}P n.m.r. spectrum (24.3 MHz at 1.4 T), what multiplicity would we predict? The signal would be split into

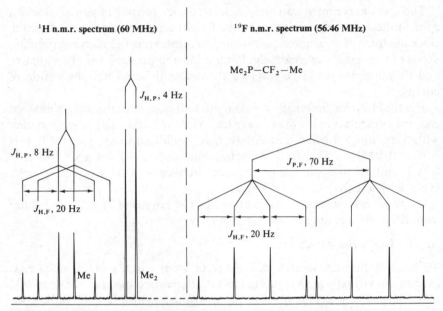

Figure 3.42   *Diagrammatic presentation of* ^{1}H *and* ^{19}F *n.m.r. spectra of* Me_2P-CF_2Me
(1.4 T).

three by the two ^{19}F nuclei (J, 70 Hz); each line would then be split into four by the single Me protons (J, 8 Hz); finally, each of these twelve lines would be split into seven by the Me_2 protons (J, 4 Hz). Considerable overlap would arise, but because of the high multiplicity the signal would be difficult to analyse.

3S.6   ELECTRON SPIN RESONANCE SPECTROSCOPY (E.S.R.)
Just as the ^{1}H nucleus has spin and therefore a magnetic moment, so the electron, with its spin, is paramagnetic and also possesses a magnetic moment. Like the proton too, the electron will precess in an applied magnetic field with a precise precessional frequency and will undergo transitions between spin states (spin orientations) if energy of the correct frequency is applied. I (or, more usually, S) for the electron is also $\frac{1}{2}$.

Measurement of electron-spin transitions is the basis of *electron spin resonance* spectroscopy (e.s.r.), also called *electron paramagnetic resonance* (e.p.r.) spectroscopy.

The magnetic moment of an unpaired electron is about 700 times that of the proton, so that the sensitivity of e.s.r. detection is very much higher than in n.m.r. (which is fortunate, since the concentrations of unpaired electrons are often correspondingly lower): e.s.r. spectra can be recorded for radical concentrations down to 10^{-4} mol dm^{-3}, irrespective of the number of nonradical species present.

The e.s.r. experiment can only detect species having unpaired electron spin, for example organic free radicals. Paired electrons cannot be detected, since the spins of an electron pair within an orbital must remain antiparallel; if one of the spins is reversed, the Pauli exclusion principle will be violated, and if both spins could be reversed there would be no net absorption of energy.

For the electron, the energy necessary to induce spin transitions in modern e.s.r. instruments is in the microwave region of the electromagnetic spectrum, with very much higher frequencies than radiofrequency. (At 1.4 T v is 3.95×10^4 MHz.) The field strength commonly used for e.s.r. work is 0.34 T, and for this field the precessional frequency of the electron is ≈ 9.5 GHz (9.5 gigahertz, 9.5×10^9 Hz).

We must now discuss a few features in the language of e.s.r. that differ from the corresponding features in n.m.r.

3S.6.1   *Derivative curves*

To begin with, e.s.r. spectra look different from n.m.r. spectra, since e.s.r. spectra are virtually always plotted as first-derivative spectra. Conventional n.m.r. signals are a plot of absorption against field strength, while e.s.r. signals are a plot of the *rate of change of absorption* against field strength; see figures 3.43 and 3.44. In general, e.s.r. absorptions are broad compared to n.m.r., and first-derivative spectra enable more accurate measurement of the spacings to be made: conversely, first-derivative spectra of n.m.r. signals would be hopelessly overlapped with one another in complex spin-coupling systems.

Another difference between e.s.r. and n.m.r. is that while a molecule may contain several ^{1}H nuclei and give rise to several ^{1}H n.m.r. signals (usually split by coupling) a radical contains only *one* unpaired electron, and therefore gives rise to only *one* signal (split by coupling).

3S.6.2   *g values*

In e.s.r., as in n.m.r., the position of the resonance signal has to be specified as a function of both field strength and frequency. In n.m.r., chemical shift positions (for example 60 MHz at 1.4 T) are expressed as field-independent units (δ or τ). In e.s.r. the resonance positions are expressed as a *g* value

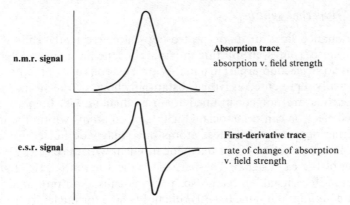

n.m.r. signal

Absorption trace
absorption v. field strength

e.s.r. signal

First-derivative trace
rate of change of absorption
v. field strength

Figure 3.43 *A n.m.r. absorption trace and an e.s.r. first-derivative trace.*

(also a function of field strength and frequency). The energy of the e.s.r. transition is given by

$$E = h\nu = gB_0 \frac{eh}{4\pi m_e c}$$

where B_0 is the external magnetic field, m_e is the electron mass, e is the electronic charge and c the speed of light: g is a proportionality constant, and has the value 2.002319 for an unbound electron. When the unpaired electron is present in an organic substrate, magnetic interactions will shift g from this value. The g values for some common organic radicals are shown below.

Note that the g values for carbon radicals are not substantially shifted from g for the unbound electron, but that oxygen and nitrogen radicals have much higher g values: this constitutes an important application to organic free-radical chemistry.

$CH_3\cdot$
2.00255

$CH_2{=}CH\cdot$
2.00220

2.0155

$C_2H_5\cdot$
2.00260

$CH_2{=}CHCH_2\cdot$
2.00254

2.00585

g values for representative radicals
(free electron $g = 2.00232$)

3S.6.3 Hyperfine splitting

The resonance lines in e.s.r. spectroscopy are frequently split in a way reminiscent of spin–spin splitting in n.m.r., but the mechanism operating in e.s.r. has no strict counterpart in n.m.r. The σ electrons in an organic molecule are normally represented as lying substantially between the atoms they bind, but electrons are not constrained to individual bonds. In particular, an unpaired electron can be associated with several atoms within the molecule in differing degrees, and if these atoms have magnetic nuclei (for example ^1H, ^{13}C) the interaction of this magnetic moment with the electron will cause splitting of the e.s.r. signal. The benzene radical anion ($C_6H_6^{-}$) gives rise to a seven-line signal, since the six protons cause splitting analogous to nuclear coupling: the same $(n + 1)$ rule holds here for nuclei with $I = \frac{1}{2}$.

This splitting in e.s.r. is called *hyperfine splitting*: the dimensions of the splitting could, as in n.m.r., be measured in Hz, but in e.s.r. the splitting is always measured in gauss ($G \equiv mT$).

The appearance of seven equally spaced lines in the e.s.r. spectrum of $C_6H_6^{-}$ is proof that all six hydrogens are equivalent, and therefore that the unpaired electron is coupling equally with all six hydrogens, rather than localised at one particular position (which would have given rise to *ortho*, *meta* and *para* splittings). Since ^{12}C is nonmagnetic no further hyperfine splitting is produced.

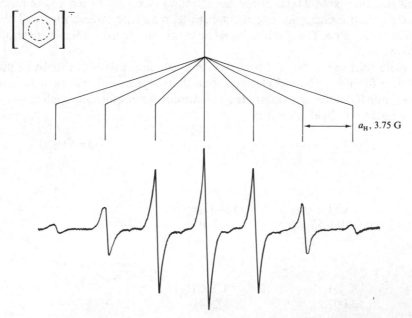

Figure 3.44   *E.S.R. spectrum of $C_6H_6^{-}$ showing the hyperfine splitting through equal coupling with six protons.*

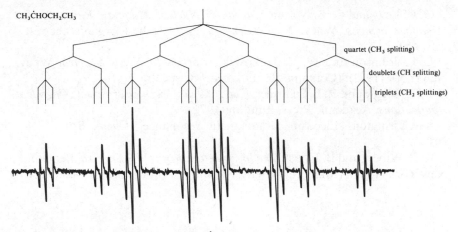

Figure 3.45 *E.S.R. spectrum of* $CH_3\dot{C}HOCH_2CH_3$. *(Radical is produced when* H^- *is abstracted from diethyl ether.)*

The electron spin density on each carbon atom of $C_6H_6^-$ is directly proportional to the observed hyperfine splitting constant a_H (3.75 G).

A second example is the e.s.r. spectrum of $CH_3\dot{C}HOCH_2CH_3$ shown in figure 3.45. The various hyperfine splittings correspond to coupling with H_3, H and H_2, in order of decreasing a_H values. The γ-CH_3 group is too distant to contribute further splitting.

FURTHER READING

MAIN TEXTS

L. M. Jackman, *Applications of Nuclear Magnetic Resonance Spectroscopy in Organic Chemistry*, Pergamon, London (1959).

J. D. Roberts, *Nuclear Magnetic Resonance*, McGraw-Hill, New York (1959).

R. H. Bible, *Interpretation of NMR Spectra*, Plenum, New York (1965).

SPECTRA CATALOGUES

NMR Spectra Catalog, Volumes 1 and 2, Varian Associates, Palo Alto, California. (Essential adjunct to a comprehensive course.)

Sadtler Research Laboratories Nuclear Magnetic Resonance Spectra, Heyden, London. (More comprehensive than the Varian catalogue.)

SUPPLEMENTARY TEXTS

NMR Quarterlies, Perkin–Elmer, Beaconsfield. (Very lucid review articles on material covered in sections 3S.1, 3S.2 and 3S.3. Series began 1971).

S. H. Pine, *J. chem. Educ.*, **49** (1972), 664. (Good review of C.I.D.N.P.)

G. C. Levy and G. L. Nelson, *Carbon-13 Nuclear Magnetic Resonance for Organic Chemists*, Wiley, New York (1972). (The best organic chemist's view so far.)

L. F. Johnson and W. C. Jankowski, *Carbon-13 NMR Spectra*, Wiley, New York (1972). (Contains 500 assigned ^{13}C spectra.)

J. H. Noggle and R. E. Schirmer, *The Nuclear Overhauser Effect: Chemical Applications*, Academic Press, London (1971).

A. Carrington, 'Electron paramagnetic resonance', *Chem. Brit.* (1968), 301.

K. B. Wiberg and B. J. Nist, *The Interpretation of NMR Spectra*, Benjamin, New York (1962). (Many fully analysed non-first order spectra included.)

4

ULTRAVIOLET AND VISIBLE SPECTROSCOPY

The absorption of ultraviolet/visible radiation by a molecule leads to transitions among the electronic energy levels of the molecule, and for this reason the alternative title *electronic spectroscopy* is often preferred.

A typical electronic absorption spectrum is shown in figure 1.4, and we can see that it consists of a series of *absorption bands*: each of these bands corresponds to an electronic transition for which $\Delta E \approx 5 \times 10^5$ J mol^{-1} or 500 kJ mol^{-1}—somewhere around the bonding energies involved in organic compounds (see sections 1.2 and 1.3).

All organic compounds absorb ultraviolet light, albeit in some instances of very short wavelength. For practical reasons, we shall be concerned with absorptions above 200 nm (see section 4.3.4), and even with this restriction we find most organic compounds show some ultraviolet absorption.

Historically, routine ultraviolet spectrometers were developed before infrared, n.m.r. or mass spectrometers, and we find now that some of the ultraviolet correlations that were previously useful have long been superseded by the later-developed techniques, so that the application of ultraviolet spectroscopy to structural investigation may appear disappointingly restricted. To take a few examples, infrared spectroscopy is the method of choice for the detection of nitrile groups or carbonyl groups (in ketones, acids, amides, etc.); n.m.r. spectroscopy will normally reveal far more information about the nature of substituents on the benzene ring than will electronic spectroscopy, etc.

The strength of electronic spectroscopy lies in its ability to measure the extent of multiple bond or aromatic conjugation within molecules. The non-

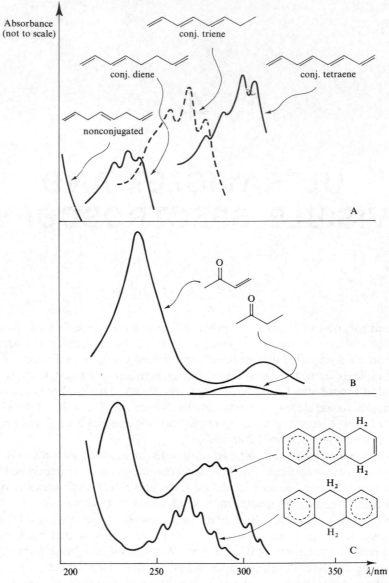

Figure 4.1 *Qualitative applications of electronic spectroscopy. (A) number of double bonds in conjugation can be determined, (B) conjugated carbonyl compounds can be distinguished from nonconjugated, (C) extent of aromatic π system can be distinguished.*

bonding electrons on oxygen, nitrogen and sulphur may also be involved in extending the conjugation of multiple-bond systems.

Electronic spectroscopy can in general differentiate conjugated dienes from nonconjugated dienes, conjugated dienes from conjugated trienes, αβ-unsaturated ketones from their βγ-analogues, etc. Since the degree of conjugation may suffer in strained molecules (examples are the loss of π-orbital overlap in 2-substituted biphenyls or acetophenones), electronic spectroscopy may be used to study such strain by correlating the change in spectrum with angular distortion. The position of absorption may also be influenced in a systematic way by substituents, and a particularly successful application has been the correlation of substituent shifts in conjugated dienes and carbonyl compounds.

The method rests heavily on empiricism, particularly in the study of aromatic and heteroaromatic systems, but here it can provide information completely unobtainable from any other spectroscopic technique.

Figure 4.1 introduces some of the qualitative aspects of electronic absorption spectroscopy: the spectra of series of isomers are presented, and the ease of distinction is noteworthy. These distinctions are achieved largely on the basis that the longer the conjugation the longer the maximum wavelength of the absorption spectrum.

4.1 COLOUR AND LIGHT ABSORPTION —THE CHROMOPHORE CONCEPT

Compounds that absorb light of wavelength between 400 and 800 nm (visible light) appear coloured to the human eye, the precise colour being a complicated function of which wavelengths the compounds subtract from white light. Very many compounds have strong ultraviolet absorption bands, the shoulders of which may tail into the visible spectrum—absorbing the violet end of the white-light spectrum. Subtraction of violet from white light leaves the complementary colours, which appear yellow/orange to the human eye and for these reasons yellow and orange are the most common colours among organic compounds. Progressive absorption from 400 nm upwards leads to progressive darkening through yellow, orange, red, green, blue, violet and ultimately black.

Chromophores. Originally, the term chromophore was applied to the system responsible for imparting colour to a compound. (The derivation is from the Greek *chromophorus*, or *colour carrier*.) Thus in azo dyes the aryl conjugated azo group (Ar—N≡N—Ar) is clearly the principal chromophore: in nitro compounds the yellow colour is carried by —NO_2, etc.

The term has been retained within an extended interpretation to imply *any functional group that absorbs electromagnetic radiation*, whether or not a 'colour' is thereby produced. Thus the carbonyl group is a chromophore both in ultraviolet and infrared terms, even though one isolated C=O group is insufficiently 'powerful' to impart colour to a compound. (An isolated carbonyl group, as in acetone, absorbs u.v. light around 280 nm.)

Important examples of organic chromophores are listed in table 4.1, and it should be stressed again that the organic application of electronic spectroscopy is mainly concerned with the conjugated chromophores at the bottom of table 4.1, to a lesser extent with those others that give rise to absorption above 200 nm, and hardly at all with those absorbing below 200 nm.

Table 4.1 Some simple organic chromophores, and the approximate wavelengths at which they absorb. The intensity of absorption is discussed in section 4.2.2.

Chromophore	Wavelength λ_{max}/nm (typical)	Intensity $\varepsilon_{max}/10^{-2}m^2\,mol^{-1}$ (typical)
C=C	175	14 000
C≡C	175	10 000
	195	2 000
	223	150
C=O	160	18 000
	185	5 000
	280	15
R—NO$_2$	200	5 000
	274	15
C≡N	165	5
C=C—C=C	217	20 000
C=C—C=O	220	10 000
	315	30
C=C—C≡C	220	7 500
	230	7 500
benzene	184	60 000
	204	7 400
	255	204

In table 4.1 the position of the maximum point of the absorption band (λ_{max}) is given. The intensity of the band at this maximum (ε_{max}) is defined in section 4.2.2.

An *auxochrome* was an earlier-defined term for a group that could enhance the colour-imparting properties of a chromophore without being itself a chromophore, examples being —OR, —NH$_2$, —NR$_2$, etc. As in the case of chromophores, the definition of auxochrome has been modified in the light of modern theory. The synergist effect of auxochromes is coupled with their ability to extend the conjugation of a chromophore by sharing of the nonbonding electrons: in a very real sense, the auxochrome then becomes part of a new, extended chromophore, and their action is only different in degree from the effect of extending the conjugation by adding a chromophore to a chromophore. There is no need to retain the term, and it will not be further used in this book.

4.2 THEORY OF ELECTRONIC SPECTROSCOPY

4.2.1 ORBITALS INVOLVED IN ELECTRONIC TRANSITIONS

When a molecule absorbs ultraviolet/visible light of a particular energy, we assume as a first approximation that only one electron is promoted to a higher energy level, and that all other electrons are unaffected. The *excited state* thus produced has a short lifetime (or the order 10^{-6} to 10^{-9} s) and a consequence is that during electronic excitation the atoms of the molecule do not move (the Franck–Condon principle).

The most probable ΔE transition would appear to involve the promotion of one electron from the highest occupied molecular orbital to the lowest available unfilled orbital, but in many cases several transitions can be observed, giving several absorption bands in the spectrum. Not all transitions from filled to unfilled orbitals are allowed, the symmetry relationship between the two orbitals being important, and this aspect is discussed further in supplement 4. Where a transition is 'forbidden', the probability of that transition occurring is low, and correspondingly the intensity of the associated absorption band is also low.

The relative energies of the molecular orbitals that most concern us are illustrated in figure 4.2. From this we can extract several generalisations.

In *alkanes*, the only transition available is the promotion of an electron from a low-lying σ orbital to a high-energy σ^* antibonding orbital: this is a high-energy process, and requires very short wavelength ultraviolet light (around 150 nm). This type of transition is classed as $\sigma \rightarrow \sigma^*$ ('sigma to sigma star').

In simple *alkenes*, several transitions are available, but the lowest energy transition is the most important: this is the $\pi \rightarrow \pi^*$ transition, which is responsible for the absorption band around 170–190 nm in unconjugated alkenes (see table 4.1).

In *saturated aliphatic ketones*, the lowest-energy transition involves the nonbonding electrons on oxygen, one of which can be promoted to the relatively low-lying π^* orbital: this n $\rightarrow \pi^*$ transition is 'forbidden' in symmetry terms (see supplement 4) and therefore the intensity is low, although the wavelength is long (around 280 nm). Two other transitions available are n $\rightarrow \sigma^*$ and $\pi \rightarrow \sigma^*$: these are both 'allowed' transitions and give rise to strong absorption bands, but the energy involved is higher than for n $\rightarrow \pi^*$; therefore the wavelength of the absorption is shorter (around 185 for n $\rightarrow \sigma^*$ and around 160 for $\pi \rightarrow \pi^*$). The most intense band for these compounds is always due to the $\pi \rightarrow \pi^*$ transition (see table 4.1).

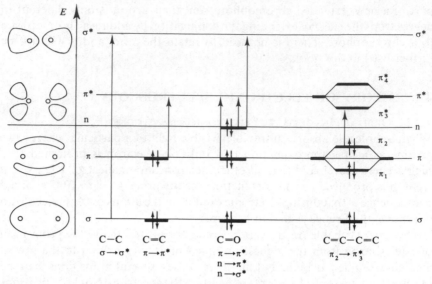

Figure 4.2   *Relative energies of orbitals most involved in electronic spectroscopy of organic compounds.*

Many factors influence the relative energies of molecular orbitals and a knowledge of these factors is the essence of an understanding of electronic spectroscopy. The influence of solvents and substituents will be discussed later, but the most dramatic effect is brought about by conjugation: the simplest case is the conjugation of two alkene groups.

In *conjugated dienes* the π orbitals of the separate alkene groups combine to form new orbitals—two bonding orbitals named π_1 and π_2 (or χ_1 and χ_2) and two antibonding orbitals named π_3^* and π_4^* (or χ_3^* and χ_4^*). The relative energies of these new orbitals are shown in figure 4.2 and it is easily apparent that a new $\pi \to \pi^*$ transition of very low energy ($\pi_2 \to \pi_3^*$) is now possible as a result of the conjugation. Conjugated dienes therefore show absorption at much longer wavelength than isolated alkene groups, a typical value being around 217 nm. Table 4.1 also shows the wavelength of absorption for a typical conjugated ketone: again the strong absorption at 220 nm is the $\pi \to \pi^*$ transition, the weaker absorption being the forbidden $n \to \pi^*$ transition.

Absorption bands appear rather than *absorption lines*, because vibrational and rotational effects are superimposed on the electronic transitions, so that an envelope of transitions arises.

The fine structure in the absorption spectra of aromatic compounds (for example benzene) is also associated with vibrational and rotational effects: in the vapour phase this fine structure is very marked, but in the liquid phase, or especially in polar solvents, the energies of the transitions are blurred out and the fine structure is less apparent.

4.2.2   LAWS OF LIGHT ABSORPTION—BEER'S AND LAMBERT'S LAWS

These two early empirical laws govern the absorption of light by molecules: Beer's law relates the absorption to the concentration of absorbing solute, and Lambert's law relates the total absorption to the optical path length.

They are most conveniently used as the combined Beer–Lambert law

$$\log(I_0/I) = \varepsilon cl \qquad \text{or} \qquad \varepsilon = A/cl$$

where I_0 is the intensity of the incident light (or the light intensity passing through a reference cell),

I is the light transmitted through the sample solution

$\log(I_0/I)$ is the *absorbance* (*A*) of the solution (formerly called the *optical density*, O.D.)

c is the concentration of solute (in mol dm^{-3})

l is the path length of the sample (in cm)

ε is the *molar absorptivity* (formerly called the *molecular extinction coefficient*).

The molar absorptivity ε is constant for a particular compound at a given wavelength, and is most commonly expressed as ε_{max}—the molar absorptivity at an absorption band maximum: typical values for ε_{max} are listed in table 4.1. Note that ε is not dimensionless, but is correctly expressed in units of 10^{-2} m^2 mol^{-1}. It is customary practice among organic chemists to omit the units (and tacitly to redefine ε) and the practice will be perpetrated in this book. Since values for ε_{max} can be very large, an alternative convention is to quote its logarithm (to the base 10), $\log_{10} \varepsilon_{max}$.

The absorbance A defined as $\log(I_0/I)$ is recorded directly by all modern double-beam instruments (see section 4.3, Instrumentation and Sampling).

If the relative molecular mass M_r (molecular weight) of a compound is unknown, and consequently ε cannot be calculated, a convenient expression is $E_{1\,cm}^{1\%}$ ('E one centimetre one per cent') defined as $E_{1\,cm}^{1\%} = A/cl$ where c in this case is the concentration in g/100 cm^3. It follows that $\varepsilon = 10^{-1}E_{1\,cm}^{1\%} \times M_r$.

4.2.3   CONVENTIONS

Absorption spectra in organic work are most often plotted with increasing ε (or log ε or $E_{1\,cm}^{1\%}$) on the ordinate, and with wavelength (λ) on the abscissa increasing from left to right. We must acknowledge that these are mere conventions: a few spectrometers plot absorbance 'upside down' to the normal convention, and, especially in physical chemistry, frequency units are sometimes used on the abscissa, corresponding to λ being plotted logarithmically and from right to left.

When a change in solvent or a change in a substituent causes λ_{max} for a band to shift, we can unambiguously state whether the shift occurs *to longer* *wavelength* or *to shorter wavelength*. (Thus, conjugation of an alkene group causes a shift to longer wavelength of the $\pi \rightarrow \pi^*$ band).

Some older established terms should be mentioned here, although their use should be progressively discouraged. *Bathochromic shift* (or *red shift*) is a shift to longer wavelength (that is, towards the red end of the spectrum). *Hypsochromic shift* (or *blue shift*) is a shift to shorter wavelength. (The expression *blue shift* is most confusing, since it could be applied to a shift from $300 \rightarrow 250$ nm, a direction that is receding from the wavelength of blue light.) A *hyperchromic effect* is one that leads to increased intensity of absorption: a *hypochromic effect* is the opposite.

4.3 INSTRUMENTATION AND SAMPLING

4.3.1 The Ultraviolet–Visible Spectrometer

In principle the modern ultraviolet–visible spectrometer is similar to the infrared spectrometer shown in figure 2.8 (page 28), consisting of a light source, double beams (reference and sample beams), a monochromator, a detector and amplification and recording devices. Several of the details differ, however: the source usually incorporates a tungsten filament lamp for wavelengths greater than 375 nm, a deuterium discharge lamp for values below that, and a solenoid-operated mirror, which automatically deflects light from either one as the machine scans through the wavelengths. The detector is usually a photomultiplier, and the ratio of reference beam to sample beam intensities (I_0/I) is fed to a pen recorder (the optical null technique is not used). The recorder trace is invariably absorbance (A) against wavelength (λ) and most instruments record linearly over the range 0 to 2 absorbance units.

4.3.2 Sample and Reference Cells

For most organic work, cells of path length 1 cm are used, although 0.1 cm and 10 cm cells are commercially available: for accurate work, the sample and reference cells should be matched in optical path length. Synthetic silica and natural silica are both used, but glass absorbs strongly from 300 nm down, and is not particularly useful for organic work.

Matched silica cells are expensive and fragile, and they must be thoroughly cleaned after each use and wiped with soft tissue dipped in methanol: they must be properly stored, and the optical surfaces must never be handled.

4.3.3 Solvents and Solutions

Electronic spectra are usually measured on very dilute solutions, and the solvent must be transparent within the wavelength range being examined. Even with double-beam operation, we cannot argue that 'solvent absorption will cancel out between the sample and reference beams' since a strongly absorbing solvent will allow so little light to pass through the cells that the photomultiplier will be effectively 'blind': this causes the amplifier to produce a very noisy background on the recorded spectrum. The lower-wavelength limits for common spectroscopic-grade solvents are shown on table 4.2: below these limits, the solvents show excessive absorbance and *sample* absorbance will not be recorded linearly. The solvents are listed in table 4.2 in an approximate order of preference, and indeed a choice of water, ethanol and hexane will meet most requirements. Spectroscopic-grade solvents are expensive and may not always be justified: for example absolute ethanol is now obtainable from ethylene hydration in the petrochemicals industry in sufficiently high purity for most spectroscopic use (the older benzene-azeotrope method always left traces of benzene in the ethanol, whereas the newer benzene–heptane azeotrope method produces ethanol with less than 2 p.p.m. of residual benzene).

Table 4.2 Solvents for spectroscopy

Solvent	Lower wavelength limit/nm
water	205
ethanol (95 per cent or absolute)	210
hexane	210
cyclohexane	210
methanol	210
diethyl ether	210
acetonitrile	210
tetrahydrofuran	220
dichloromethane	235
chloroform	245
carbon tetrachloride	265
benzene	280

Solvent shifts may occur, and comparisons between spectra should only be made with this realisation (see section 4.4).

Solution preparation should always be done with great care: standard solutions are prepared in volumetric flasks, the concentration usually being around 0.1 per cent. This concentration can be used to record the absorbances (and ε) of the weak bands, but the solution may have to be diluted to bring the intense bands on scale: if dilution is necessary, it should always be done by a known factor (for example dilute 2 cm^3 to 20 cm^3) so that ε for strong bands can also be calculated. Some compounds do not obey Beer's law exactly, so that the spectra recorded at differing concentrations do not match. This is usually a minor effect, unless the compound ionises to an increased extent at high dilutions.

It is worth stressing that modern ultraviolet/visible spectrometers are extremely accurate in the measurement of absorbance, and a more significant source of error is in our ability to prepare standard solutions at low concentrations. For accurate work, all standard flasks, etc. should be of the highest analytical quality, and if dilution of the original solution is necessary it should be carried out on a volume that can be measured with the required accuracy; too small a volume will certainly lead to dilution errors.

Replotting of the spectrum should always be done to convert the absorbance spectrum (from the spectrometer) into the ε or log ε form: this involves calculating ε from concentration and absorbance figures and manually redrawing the spectrum on graph paper.

4.3.4   VACUUM ULTRAVIOLET

The conventional electronic spectroscopic techniques outlined above cannot be used below 200 nm since oxygen (in the air) begins to absorb strongly there. To study the higher-energy transitions below 200 nm (see table 4.1) the entire path length must be evacuated, and for this reason the region below 200 nm is usually referred to as the *vacuum ultraviolet*. The equipment for this work is more expensive, and a more highly trained technique is demanded to operate it: it has mainly been exploited in studying bond energies, etc., and is not usually very helpful in organic structural determinations.

4.4   SOLVENT EFFECTS

The position and intensity of an absorption band may shift when the spectrum is recorded in different solvents. For changes to solvents of *increased polarity* we can summarise the normal pattern of shifts as follows.

Conjugated dienes and *aromatic hydrocarbons* experience very little solvent shift.

$\alpha\beta$-*Unsaturated carbonyl compounds* show two different shifts: (a) the $\pi \rightarrow \pi^*$ *band* moves to longer wavelength (red shift) while (b) the $n \rightarrow \pi^*$ *band* moves to shorter wavelength (blue shift).

We can express this general picture in the form of an energy diagram as in figure 4.3; solvation by a polar solvent stabilises π, π^* and n orbitals: the stabilisation of nonbonding orbitals is particularly pronounced with hydrogen-bonding solvents (such as water or ethanol), and π^* orbitals are more stabilised by solvation than π orbitals, presumably because π^* orbitals are the more polar. The net result is as shown in figure 4.3: the energy of transition $\pi \rightarrow \pi^*$ becomes less with solvation (red shift) while the energy of transition $n \rightarrow \pi^*$ becomes greater (blue shift).

The dimensions of the shift in $\alpha\beta$-unsaturated ketones are listed in table 4.4: the maximum shift (19 nm) occurs in the $\pi \rightarrow \pi^*$ band on changing the solvent from hexane or cyclohexane to water.

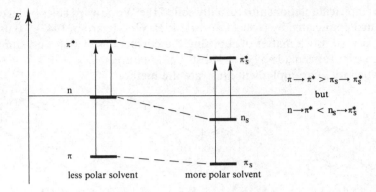

Figure 4.3 *Effect of solvation on the relative energies of orbitals and transitions in αβ-unsaturated carbonyl compounds. In more polar solvents π → π* absorptions move to longer λ, n → π* move to shorter λ.*

4.5 APPLICATIONS OF ELECTRONIC SPECTROSCOPY — CONJUGATED DIENES, TRIENES AND POLYENES

Figure 4.1 shows the characteristic progression in the electronic spectra of dienes, trienes and tetraenes: the progression continues with increased conjugation up to an asymptotic limit of around 450 nm (11 or 12 double bonds in conjugation) by which stage the polyenes are strongly yellow in colour. The red colour of tomatoes and carrots arises from conjugated molecules of this type.

For dienes and trienes the position of the most intense band can be correlated in most instances with the substituents present: table 4.3 summarises

Table 4.3 Conjugated dienes and trienes (in ethanol) λ_{max} for π → π* transitions. ε_{max} 6000–35 000 ($\times 10^{-2} m^2 mol^{-1}$)

acyclic and heteroannular dienes	215 nm
homoannular dienes	253 nm
acyclic trienes	245 nm

Addition for each substituent

—R alkyl (including part of a carbocyclic ring)	5 nm
—OR alkoxy	6 nm
—SR thioether	30 nm
—Cl, —Br	5 nm
—OCOR acyloxy	0 nm
—CH=CH— additional conjugation	30 nm
if one double bond is exocyclic to one ring	5 nm
if exocyclic to two rings simultaneously	10 nm

solvent shifts minimal

these empirical relationships (usually called the Woodward rules—since they were first enunciated by Nobel Laureate R.B. Woodward in 1941). To use the rules, it is simply a matter of choosing the correct base value, and summing the contributions made by each substituent feature.

Two worked examples will illustrate the method.

I

II

III

IV

For compound I the base value is 214 nm, since the two double bonds are heteroannular: there are 4 alkyl substituents (the ring residues a, b, c, and the methyl group d), adding 4 × 5 nm: the double bond in ring A is exocyclic to ring B, adding 5 nm: the total is 214 + 20 + 5 = 239 nm. This is within 2 nm of the observed value.

For compound II the base value is 253 nm, to which we add 4 × 5 (four ring residues a, b, c, and d) and 2 × 5 (both double bonds are exocyclic, to rings A and C, respectively) giving a total of 283 nm. This is within 2 nm of the observed value.

Compounds III and IV should be similarly treated, as a drill exercise.

Further examples of the application of Woodward's rules are given in chapter 6.

It is worth restating that these rules are empirically based on analogous model compounds, and will only hold good for other compounds whose structures are close to the models. The influence of contradictory factors is discussed in section 4.11.

It is not possible to predict ε_{max} other than within the range given in table 4.3: the only general rule is that longer conjugation leads to higher intensity, the *approximate* relationship in many cases being $\varepsilon \propto$ (length of chromophore)^2.

4.6 APPLICATIONS OF ELECTRONIC SPECTROSCOPY —CONJUGATED POLY-YNES AND ENEYNES

Interest in these compounds arose from their discovery in plant material, and electronic spectroscopy has been of paramount importance in structure elucidation: the number of possible permutations between triple and double bonds in polyeneynes is very large indeed, but two typical groupings are shown below.

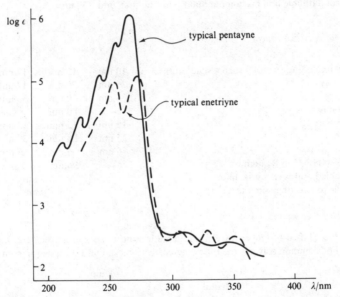

The absorption spectrum of a typical poly-yne and a typical polyeneyne are shown in figure 4.4: as expected, the intensity and λ_{max} values increase with increasing conjugation and substitution, so that assistance in proof of structure of a new compound can be achieved by comparison of the new compound's spectrum with that of a closely related model compound.

Figure 4.4 *Electronic absorption spectra typical of poly-yne and polyeneyne chromophores.*

4.7 APPLICATIONS OF ELECTRONIC SPECTROSCOPY —αβ-UNSATURATED CARBONYL COMPOUNDS

Table 4.4 shows the correlations that can apply to αβ-unsaturated carbonyl compounds: the original data were compiled by Woodward, with many extensions and verifications being added over the years.

The method of use is similar to that for conjugated dienes (table 4.3), the only two variants being (a) the distinction between α and β substituents and (b) the importance of solvent shifts (see also section 4.4).

Table 4.4　$\alpha\beta$-Unsaturated carbonyl compounds (in ethanol)
λ_{max} for $\pi \to \pi^*$ transitions. ε_{max} 4500–20 000 ($\times\ 10^{-2} m^2\ mol^{-1}$)

ketones $-\overset{\beta}{\underset{\vert}{C}}=\overset{\alpha}{\underset{\vert}{C}}-CO-$ acyclic or 6-ring cyclic	215 nm
5-ring cyclic	202 nm
aldehydes $-\underset{\vert}{C}=\underset{\vert}{C}-CHO$	207 nm
acids and esters $-\underset{\vert}{C}=\underset{\vert}{C}-CO_2H(R)$	197 nm

Additional conjugation

$-\overset{\delta}{\underset{\vert}{C}}=\overset{\gamma}{\underset{\vert}{C}}-\overset{\beta}{\underset{\vert}{C}}=\overset{\alpha}{\underset{\vert}{C}}-CO-$ etc.	add　30 nm
(If the second double bond is homoannular with the first, add	39 nm)

Addition for each substituent

	α	β	γ	δ
—R alkyl (including part of a carbocyclic ring)	10 nm	12 nm	17 nm	17 nm
—OR alkoxy	35 nm	30 nm	17 nm	31 nm
—OH hydroxy	35 nm	30 nm	30 nm	50 nm
—SR thioether	—	80 nm	—	—
—Cl chloro	15 nm	12 nm	12 nm	12 nm
—Br bromo	25 nm	30 nm	25 nm	25 nm
—OCOR acyloxy	6 nm	6 nm	6 nm	6 nm
—NH$_2$, —NHR, —NR$_2$ amino	—	95 nm	—	—
if one double bond is exocyclic to one ring		5 nm		
If exocyclic to two rings simultaneously		10 nm		

Solvent shifts (see section 4.4)

Above values shifted to longer wavelength in water, and to shorter wavelength in 'less polar' solvents. For common solvents the following corrections should be applied in computing λ_{max}

water	+8 nm
methanol	0
chloroform	−1 nm
dioxan	−5 nm
diethyl ether	−7 nm
hexane	−11 nm
cyclohexane	−11 nm

The typical appearance of an $\alpha\beta$-unsaturated-ketone spectrum is shown in figure 4.1: note that these Woodward rules calculate the position of the intense $\pi \to \pi^*$ transition, not the weak n $\to \pi^*$ transition. As for conjugated dienes, ε_{max} cannot be predicted with accuracy.

Two worked examples will illustrate the method.

For compound I, the base value is 215 nm: for one α-alkyl group add 10 nm, for one β-alkyl group add 12 nm; the total is 215 + 10 + 12 = 237 nm.

3-methylpent-3-en-2-one

I   II   III   IV

This is within 1 nm of the observed value (for which ε_{max} is 4600). Had the spectrum been recorded in water, λ_{max} would have moved to 245 nm.

For compound II, the base value is again 215 nm: for one α-alkyl group, add 10 nm: for two β-alkyl groups add 2×12 nm: for a double bond exocyclic to two rings add 10 nm: the total is $215 + 10 + 24 + 10 = 259$ nm. This is within 3 nm of the observed value (for which ε_{max} is 6500). Had the spectrum been recorded in hexane, λ_{max} would have moved to 248 nm.

Compounds III and IV should be similarly treated as a drill exercise.

In general, the same reservations apply to the predictions of $\alpha\beta$-unsaturated ketone absorptions as apply to conjugated dienes (see section 4.5).

4.8 APPLICATIONS OF ELECTRONIC SPECTROSCOPY —BENZENE AND ITS SUBSTITUTION DERIVATIVES

It has already been stated (in section 4.2.1) that the electronic absorption spectrum of benzene shows a great deal of fine structure in the vapour phase; less is seen in hexane solution, and less still in ethanol solution.

A similar trend takes place when benzene is substituted—simple alkyl substituents shift the absorptions slightly to longer wavelength, but do not destroy the fine structure, while nonbonding pair substituents (—OH, —OR, —NH$_2$, etc.) shift the absorptions more substantially to longer wavelength and seriously diminish (or wholly eliminate) the fine structure.

It is only very occasionally necessary to rely on the electronic spectrum to infer structural relationships in substituted benzenes: additionally, the empirical rules that have been compiled for calculating their λ_{max} values are often ambiguous, and do not permit predictions concerning the number of bands present nor their intensity. Given the enormous number of possible substitution patterns, the only consistently valid approach to predicting the electronic spectrum of anything but the simplest benzene derivative is to record the electronic spectrum of a suitable model compound or to find such a spectrum in the literature (see Further Reading). It would be unwise to extrapolate substantially from the model, but the following indicators are worth recording.

1. The most 'influential' combination of substituents is a $-$M group *para* to a $+$M group, so that nonbonding pair donation (by the $+$M group) is

effectively complementary to the electron-withdrawing $-M$ group: the shift in λ_{max} is greater than the sum of the individual shifts. (Examples are *p*-nitroaniline and *p*-nitrophenol.)

2. Usually a $-M$ group *ortho* or *meta* to a $+M$ group produces merely a small shift from that of the isolated chromophores.

3. Altering the nonbonding pair 'availability' will alter the position of λ_{max}. (Examples are the pronounced red shift when *p*-nitrophenol is converted by base to the *p*-nitrophenate ion, or the blue shift caused by protonation of amino groups.)

4.9   APPLICATIONS OF ELECTRONIC SPECTROSCOPY —AROMATIC HYDROCARBONS OTHER THAN BENZENE

Figure 4.5 shows the characteristic development of ultraviolet–visible absorption as benzene rings are fused together in the linear series benzene–naphthalene–anthracene or angularly in the series benzene–naphthalene–

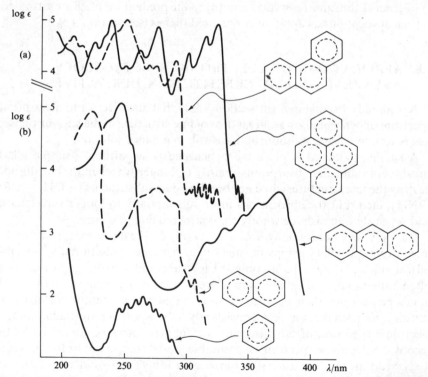

Figure 4.5   *Electronic-absorption spectra of typical polynuclear aromatic-hydrocarbon chromophores. All spectra recorded in hexane. For clarity the upper set (a) (phenanthrene and pyrene) has been displaced upwards on the ordinate from set (b) (benzene, naphthalene and anthracene). The wavelength scale for both sets is identical.*

phenanthrene. The fusion of additional rings leads to more and more complex spectra (see that of pyrene in figure 4.5), *but these spectra are uniquely characteristic of each aromatic chromophore*, so that it is a very simple matter to compile a spectra catalogue of aromatic hydrocarbons for identification purposes. The introduction of alkyl groups has little influence on λ_{max} or ε_{max}, so that (for example) methylanthracenes are readily identified as possessing the anthracene chromophore from their electronic spectra, etc.

Table 4.5 lists the λ_{max} and ε_{max} values for many common hydrocarbon systems.

4.10 APPLICATIONS OF ELECTRONIC SPECTROSCOPY —HETEROCYCLIC SYSTEMS

Once again, the most successful approach to the electronic spectra of heterocyclic systems has been an empirical one, coupled with a few guidelines on substituent effects.

Table 4.6 lists the principal λ_{max} and ε_{max} values for common heterocyclic systems.

Simple alkyl substituents as usual have little effect on the spectra, but polar groups (electron donors or attractors) can have profound effects, which are usually highly dependent on substitution position in relation to the heteroatom(s). The possibility that tautomeric systems may be generated should also be borne in mind, the classic case here being the 2-hydroxypyridines, which tautomerise almost entirely to 2-pyridones, with substantial changes in the electronic spectra.

4.11 STEREOCHEMICAL FACTORS IN ELECTRONIC SPECTROSCOPY

The empirical rules met until now are much honoured in the breach, a variety of reasons being responsible for the failures: consistently, stereochemical reasons are to be blamed and it is worth taking a few examples to illustrate this. In all cases it can be shown that angular strain or steric overcrowding has distorted the geometry of the chromophore, so that, for example, conjugation is reduced by reducing the π-orbital overlap, etc.

4.11.1 BIPHENYLS AND BINAPHTHYLS

Biphenyl (I) is not completely planar (the two rings being at an angle of approximately 45°) and in 2-substituted biphenyls (II) the two rings are pushed even further out of coplanarity: the result is that diminished π-orbital overlap in the 2-substituted derivatives leads to blue shifts and diminished intensity in their electronic spectra.

Thus biphenyl (I) has λ_{max} 250 (ε, 19 000) while 2-methylbiphenyl (II) has λ_{max} 237 (ε, 10 250). Adding more methyl groups is complicated by the bathochromic effect of the methyl groups themselves, but an interesting

Table 4.5 Principal maxima in the electronic spectra of aromatic hydrocarbons†

Hydrocarbon	Solvent	Principal maxima
benzene	E	229 (1.21), 234 (1.46), 239 (1.76), 243 (2.00), 249 (2. 30), 254 (2.36), 260 (2.30), 268 (1.04)
toluene, xylenes, and trimethylbenzenes		similar appearance to benzene spectrum; but peaks move to longer λ and higher ε with each additional alkyl group
biphenyl	E	250 (4.15); 2- or 2′-substituents may change the spectrum considerably
binaphthyls	H	220 (5.00), 280 (4.20)‡; see biphenyl
indene	H	209 (4.34), 221 (4.03), 249 (3.99), 280 (2.69), 286 (2.35); many inflections
styrene	H	247 (4.18), 273 (2.88), 282 (2.87), 291 (2.76)
fluorene	E	261 (4.23), 289 (3.75), 301 (3.99)
naphthalene	E	221 (5.00), 248 (3.40), 266 (3.75), 275 (3.82), 285 (3.66), 297 (2.66), 311 (2.48), 319 (1.36)
acenaphthene	E	similar appearance to naphthalene spectrum
trans-stilbene	E	230 (4.20), 299 (4.48), 312 (4.47)
cis-stilbene	E	225 (4.34), 282 (4.10)
phenanthrene	E	223 (4.25), 242 (4.68), 251 (4.78), 274 (4.18), 281 (4.14), 293 (4.30), 309 (2.40), 314 (2.48), 323 (2.54), 330 (2.52), 337 (3.40), 345 (3.46)
phenalene (perinaphthene)	E	234 (4.40), 320 (3.90), 348 (3.70)
fluoranthene	E	236 (4.66), 276 (4.40), 287 (4.66), 309 (3.56), 323 (3.76), 342 (3.90), 359 (3.95)
chrysene	E	220 (4.56), 259 (5.00), 267 (5.20), 283 (4.14), 295 (4.13), 306 (4.19), 319 (4.19), 344 (2.88), 351 (2.62), 360 (3.00)
pyrene	E	231 (4.62), 241 (4.90), 251 (4.04), 262 (4.40), 272 (4.67), 292 (3.62), 305 (4.06), 318 (4.47), 334 (4.71), 352 (2.82), 362 (2.60), 372 (2.40)
anthracene	E	252 (5.29), 308 (3.15), 323 (3.47), 338 (3.75), 355 (3.86), 375 (3.87)
perylene	E	245 (4.44), 251 (4.70), 387 (4.08), 406 (4.42), 434 (4.56)
acenaphthylene	H	229 (4.72), 264 (3.46), 274 (3.43), 311 (3.93), 323 (4.03), 334 (3.70), 340 (3.70), 440 (2.00), 468 (1.56)

† Values quoted are for λ_{max} in nm, with log ε_{max} in parentheses. Solvents used were either hexane (H) or ethanol (E).
‡ 2,2′-binaphthyls are different: 255 (4.9), 320 (4.4).

With permission from *Qualitative Organic Analysis* by W. Kemp, McGraw-Hill, Maidenhead (1970).

comparison can be made between the hexamethylbiphenyl (III) and mesitylene (1,3,5-trimethylbenzene, IV): these two exhibit the same λ_{max} (266 nm) and their ε_{max} values are 545 and 260, respectively (in ethanol).

The 1,1′-binaphthyls cannot be coplanar (and indeed are enantiomeric) and exhibit electronic absorption at much shorter wavelength than for naphthalene itself (see table 4.5): in 2,2′-binaphthyls, with much less over-

Table 4.6 Principal maxima in the electronic spectra of some common heterocyclic systems†

Compound	Solvent	Principal maxima
pyrrole	E	235 (2.7, shoulder); no sharp maxima
furan	H	207 (3.96)
thiophen	H	227 (3.83), 231 (3.85), 237 (3.82), 243 (3.58)
indole	H	220 (4.42), 262 (3.80), 280 (3.75), 288 (3.61)
carbazole	E	234 (4.63), 244 (4.38), 257 (4.29), 293 (4.24), 324 (3.55), 337 (3.50)
pyridine	H	251 (3.30), 256 (3.28), 264 (3.17)
quinoline	E	226 (4.53), 230 (4.47), 281 (3.56), 301 (3.52), 308 (3.59)
isoquinoline	H	216 (4.91), 266 (3.62), 306 (3.35), 318 (3.56)
acridine	E	249 (5.22), 351 (4.00)
pyridazine	H	241 (3.02), 246 (3.15), 251 (3.15), 340 (2.56)
pyrimidine	H	242 (3.31), 293 (2.51), 307 (2.40), 313 (2.18), 317 (2.04), 324 (1.73)
pyrazine	H	254 (3.73), 260 (3.78), 267 (3.57), 315 (2.93), 322 (2.99), 328 (3.02)
barbituric acids	water	256–7 ($\approx$4.4, concentration dependent)

† Values quoted are for λ_{max} in nm, with log ε_{max} in parentheses. Solvents used were usually hexane (H) or ethanol (E); change of solvent may affect the spectrum considerably.

With permission from *Qualitative Organic Analysis* by W. Kemp, McGraw-Hill, Maidenhead (1970).

I   II   III   IV

crowding, the λ_{max} values are nearer those expected from the benzene–biphenyl analogy.

4.11.2 *cis* AND *trans* ISOMERS

Where an alkene chromophore is capable of geometrical isomerism, it is normally found that the *trans* isomer exhibits longer-wavelength absorption (and higher intensity) than the *cis* isomer: this can be rationalised in terms of the more effective π-orbital overlap possible in the *trans* isomer, and is illustrated well in the case of *cis* and *trans*-stilbenes (see table 4.5).

4.11.3   ANGULAR DISTORTION AND CROSS-CONJUGATION. STERIC INHIBITION OF RESONANCE

The Woodward rules for conjugated dienes and carbonyl compounds give reliable results only where there is an absence of strain around the chromophore: thus the rules are successful for acyclic and (most) six-membered ring systems. Well-authenticated violations are abundant, whether the chromophore distortion is engendered by ring strain or by the introduction of additional conjugation other than at the end of the chromophore (cross-conjugation). Examples of systems whose electronic spectra could not satisfactorily be predicted by the Woodward rules are shown below.

I   II   III

IV   V   VI

So sensitive is electronic spectroscopy to distortion of the chromophore, that one can turn this to advantage in *demonstrating* that distortion is present in the molecule. In molecules IV, V, and VI, the observed λ_{max} values are lower than calculated, thus demonstrating the influence of steric inhibition of resonance in the conjugation; further examples of steric inhibition of resonance arise in substituted biphenyls and binaphthyls (see section 4.11.1).

SUPPLEMENT 4

4S.1   QUANTITATIVE ELECTRONIC SPECTROSCOPY

The high sensitivity of electronic spectroscopy and the ease and accuracy with which quantitative work can be carried out, make it a valuable analytical method—limited principally by the need for a chromophore in the system under analysis.

From the Beer–Lambert law (see section 4.2.2), expressed in the form $\varepsilon = A/cl$, we can see that the concentration of a chromophoric molecule can be measured in solution provided we know the value of ε: the path length (l) is the cell dimension, and absorbance (A) is obtained as $\log(I_0/I)$ from the spectrometer.

A simple example of an analytical problem would be the estimation of a mixture of anthracene and naphthalene. Figure 4.5 shows that in ethanol anthracene absorbs at λ_{max} 375 nm (log ε, 3.87), well clear of any naphthalene absorption. We can prepare a standard solution of the anthracene–naphthalene mixture and measure the absorbance at 375 nm: from the Beer–Lambert law the concentration of pure anthracene can be calculated and hence the proportion of anthracene in the mixture.

Direct measurement of the naphthalene concentration is complicated by the fact that, at all λ_{max} for naphthalene, part of the absorbance will be due to a contribution from anthracene. The procedure would be to choose a suitable λ_{max} for naphthalene (for example λ_{max} 285 nm), and measure *on the anthracene spectrum the value of ε for anthracene* at this wavelength. (From figure 4.5, log ε at 285 nm for anthracene is 2.2). Since we already know the concentration of anthracene, and now also ε at 285 nm, we can *calculate* the absorbance due to anthracene (A_a) at 285 nm (for a given mixture solution). We can *measure* the total absorbance at 285 nm for the same solution (A_t) and hence calculate the absorbance due to naphthalene at 285 nm (A_n) since $A_n = A_t - A_a$. From A_n we can then calculate the concentration of naphthalene in the solution and hence the proportion of naphthalene in the mixture.

This latter technique must be used if (say) anthracene and naphthalene are present in a mixture with other nonabsorbing constituents, so that the proportion of naphthalene cannot be calculated simply by difference.

Simple extensions of the two basic analyses outlined above, with the ever-desirable compiling of calibration curves for more complex cases, have been used to meet the vast majority of quantitative applications.

4S.2 FLUORESCENCE AND PHOSPHORESCENCE

While anthracene is 'colourless', in the sense that its electronic absorption spectrum lies wholly within the ultraviolet region (see table 4.5), pure samples of anthracene viewed in ultraviolet light give off a blue visible light, which we call fluorescence. *Fluorescence is light that is emitted from a molecule after the molecule has absorbed light of a different (and shorter) wavelength.*

One characteristic feature of fluorescent radiation is that the fluorescence stops whenever the irradiating light is removed. A related phenomenon, phosphorescence, arises when molecules continue to emit the longer-wavelength radiation even after the exciting radiation has been removed.

The explanation of fluorescence and phosphorescence can best be viewed from a study of figure 4.6: this represents the one-electron excitation process discussed earlier (in section 4.2) but introduces additional nomenclature and two additional considerations.

The absorption of light hv_A by the ground-state molecule leads to promotion of one electron from a ground-state M.O. to one of the vibrational sublevels of the first excited-state M.O., S_1 (or to some higher electronic excited state S_2, S_3, etc.). This rapidly deactivates in solution, by a radiationless process, to the *lowest vibrational sublevel of* S_1.

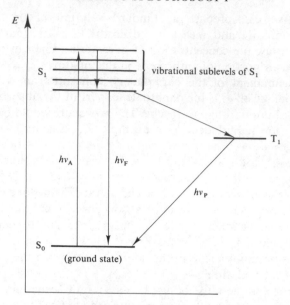

Figure 4.6 *Mechanism of fluorescence and phosphorescence. The electronic transitions hv_A, hv_F and hv_P are, respectively, absorption, fluorescence and phosphorescence.*

For the electron to return from S_1 to the ground state, it must emit radiation (hv_F), which is clearly of lower energy and thus of longer wavelength than hv_A: hv_F is the *fluorescent radiation*.

An alternative path involves loss of energy from S_1 to T_1: the principal difference between S_1 and T_1 is the electron spin orientation. For the two electrons originally occupying the ground-state M.O. under consideration, their spins must be antiparallel following the Pauli exclusion principle: the original excitation, followed by the decay to the lowest vibrational sublevel of S_1, does not alter the spin of the promoted electron, but the transition S_1 to T_1 does. Energy states containing only spin-paired electrons are called *singlet states* (S_0, S_1, S_2) while those with parallel spin electrons are called *triplet states* (T_1, etc.). Triplet states are more stable than singlet states, and are longer lived: they may survive after the exciting radiation has been removed, and *thereafter* decay to S_0 by emitting the radiation hv_P, the *phosphorescent radiation*.

For complex molecules like anthracene, several fluorescent bands are emitted, constituting the fluorescence spectrum, which is therefore, in origin, not unlike the Raman spectrum discussed in section 2S.3: anthracene will emit this same fluorescence spectrum even though the wavelength of the excitation is varied within quite wide limits (although the intensity of the fluorescence spectrum will vary). The fluorescence spectrum for anthracene consists of four maxima, with λ_{max} 380, 400, 420 and 450 nm: this brings the emission into the blue region of the visible spectrum, and therefore anthracene fluoresces blue.

Important uses of the fluorescence phenomenon are represented by the molecules of fluorescein and the 'optical brightener' derivative of 4,4'-diaminostilbene (also called a 'fluorescent brightening agent' or 'optical bleach').

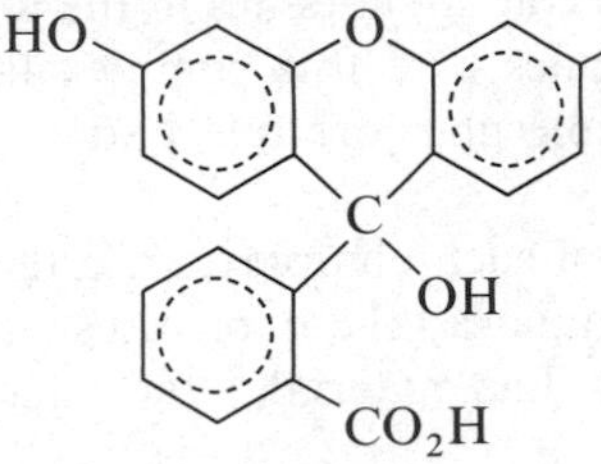

fluorescein
(acid form)

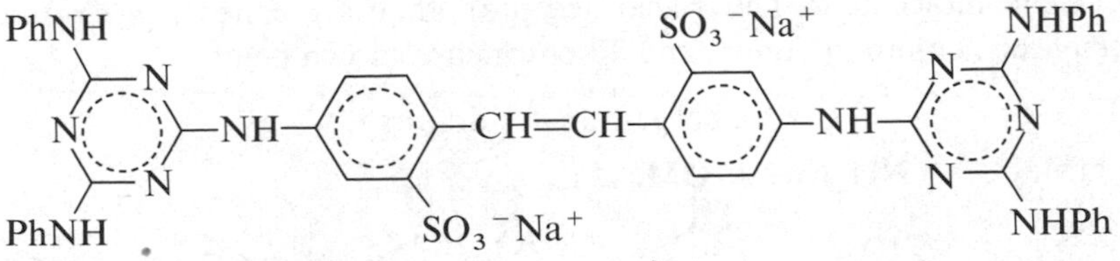

an optical brightener

The intense green fluorescence of aqueous fluorescein solutions makes it an excellent material to add to water systems for leak detection, and an excellent 'marker' for sea rescue operations, etc.

Minute quantities of optical brighteners are added to detergents, and are retained on the fabrics after washing: in sunlight (or other ultraviolet source) they fluoresce blue and add brightness (and whiteness?) to the fabric.

Other applications of fluorescence depend on its extremely low detection limit: it is used in polymer chemistry to detect and identify plasticisers, and in the study of impurities (for example oxidation impurities in poly-ethylene and polypropylene). Fluorescent material dissolved in solution or in solid plastic bases can also detect radioactive decay: this is the foundation of *scintillation counting* of β-emitters, etc. Biological applications include the study of the 3-dimensional tertiary structure of proteins by measuring the proximity of known fluorescent groups within the protein: these fluorescent groups can be either aromatic amino acids, or specifically added fluorescent labels.

(It might be interesting to contrast that the 'phosphorescence' seen at night in the sea is due to several species of marine micro-organism, which, when agitated by a bow-wave or an oar-splash, undergo an enzymatic alarm reaction which liberates energy in the form of green light.) This phenomenon is more correctly called *bioluminescence*.

4S.3 ABSORPTION SPECTRA OF CHARGE-TRANSFER COMPLEXES

Solutions of iodine in hexane are violet in colour, while in benzene they are brown: if tetracyanoethylene TCNE (colourless) is added to a chloroform solution of aniline (colourless) the result is a deep blue coloured solution.

The explanation for these colour shifts lies in the formation of complexes between the pairs of molecules, and this complex formation leads to the production of two new molecular orbitals and consequently to a new electronic transition.

Perhaps the best known of such complexes are the picrates of aromatic hydrocarbons, esters and amines; these picrates are usually sufficiently stable to be isolable as crystalline material, although some picrates are only stable in solution.

Since the formation of these complexes involves transfer of electronic charge from an 'electron-rich' molecule (a Lewis-base donor) to an 'electron-deficient' molecule (a Lewis-acid acceptor) they are called *charge-transfer complexes*: common donors and acceptors are shown below.

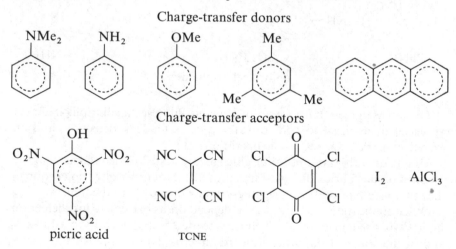

Charge-transfer donors

Charge-transfer acceptors

picric acid   TCNE

Bond formation between the molecular pairs is brought about when filled π-orbitals (or nonbonding orbitals) in the donor overlap with depleted orbitals in the acceptor. The two new molecular orbitals formed are illustrated in figure 4.7: the lower-energy M.O. for the complex is occupied in the ground state, and transitions from this M.O. to the new upper M.O. are responsible for the new absorption bands formed.

The structure of most charge-transfer complexes can be visualised as a face-to-face association on a 1:1 donor:acceptor basis: only thus, for example, can maximum overlap of aromatic π-orbitals take place. This kind of

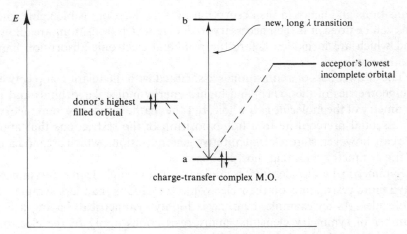

Figure 4.7 *Electronic transitions for charge-transfer complexes. Donor and acceptor orbitals combine to form two new orbitals (a and b) for the complex. New electronic transitions for long λ are then possible. between a and b.*

structure is difficult to draw, and most representations use one or other of the conventions shown below.

donor acceptor   donor
(anthracene picrate)   acceptor

From the large number of possible donors and acceptors, a very large number of charge-transfer complexes can be formed, and it would be impossible to discuss their electronic-absorption spectra. The benzene–iodine and aniline–TCNE cases are representative; λ_{max} values for aniline and TCNE are 280 nm and 300 nm, respectively, while the deep blue complex has λ_{max} well into the visible at 610 nm. In the benzene–iodine case, λ_{max} for benzene is 255 nm, while for molecular iodine in hexane λ_{max} is in the visible around 500 nm: the charge-transfer complex has an intense additional band around 300 nm, but this tails into the visible region and modifies the violet colour of the molecular iodine to brown.

4S.4 SYMMETRY RESTRICTIONS ON THE ALLOWEDNESS OF ELECTRONIC TRANSITIONS

Even for simple organic molecules it is generally difficult to calculate the energies of the molecular orbitals that might be involved in electronic transitions and hence in the absorption of ultraviolet or visible light: where this can be done, it is then necessary to decide which transitions are allowed and which are forbidden before the principal electronic absorption bands can be predicted.

The allowedness of a transition is associated with the relationship between the geometries of the lower- and higher-energy molecular orbitals and the symmetry of the molecule as a whole, but the mathematics of group theory is an essential prerequisite to a full exposition of the restrictions that apply. We can, however, state a few qualitative generalisations, which are indicators of the symmetry rules that must be observed.

Symmetrical molecules (that is, molecules with a high degree of symmetry) have more restrictions on their electronic transitions than less symmetrical molecules. As an example, benzene is highly symmetrical, having a large number of symmetry elements: many restrictions apply to the electronic transitions of the benzene molecule and therefore its electronic absorption spectrum is simple.

For a *totally* unsymmetrical molecule, the only symmetry operation that can be performed is the identity operation (which changes nothing). For such a molecule, no symmetry restrictions apply to the electronic transitions, so that transitions may be observed among *all* of its molecular orbitals except, of course, among filled orbitals: a complex electronic absorption spectrum will result.

Between these two extremes lies the majority of organic compounds that absorb light in the ultraviolet–visible region. For any of these compounds, to decide whether a transition between two given molecular orbitals is allowed or forbidden, and will give rise to ultraviolet–visible absorption, we must consider (a) the geometry of the ground-state M.O. (b) the geometry of the excited-state M.O. and (c) the orientation of the electric dipole of the incident light that might induce the transition (when referred to the same co-ordinates). Provided these three have an appropriate symmetry relationship, the transition will be allowed.

4S.5 OPTICAL ROTATORY DISPERSION AND CIRCULAR DICHROISM

4S.5.1 Definitions and nomenclature

The optical activity of a compound is its ability to rotate the plane of polarised light, and we define the *specific rotation*, $[\alpha]$, as $[\alpha] = 100\alpha/lc$, where α is the observed rotation, l is the length of polarimeter tube (in dm) and c the concentration (in g/100 cm^3). Since $[\alpha]$ may vary with temperature and with the wavelength of light used, these two must also be specified; routine measurements use the sodium D-line at 589 nm, so that for data at 20°C we would quote specific rotation as $[\alpha]_D^{20}$. The solvent used must also be quoted, and for accurate work it is better also to quote the concentration used.

The fact that $[\alpha]$ varies with wavelength is an important one: the phenomenon is named *optical rotatory dispersion* (O.R.D.), and O.R.D. curves play an important role in structure elucidation in optically active compounds.

Figure 4.8 shows that we can regard plane polarised light as the resultant of two equal and opposite beams of circularly polarised light, and this leads to two derivations.

(i) If one of the circularly polarised components passes through the medium more slowly than the other, *the plane of polarisation will be rotated.* (For this to arise, the refractive index of the medium for one circularly polarised component must be different from that for the other, the phenomenon being named *circular birefringence.*)

(ii) Whenever we find circular birefringence, we also find that the two circularly polarised components are differentially absorbed by the medium: this leads to the emergent beam having an imbalance between the strengths of the two circularly polarised beams, so that the emergent beam is not truly plane polarised but *elliptically* polarised (see figure 4.8). We name this phenomenon *circular dichroism* (C.D.).

The combined phenomena of circular birefringence and circular dichroism are named the *Cotton effect.*

4S.5.2   Cotton effect and stereochemistry

The O.R.D. curves for many compounds show typically the *plain curve* variance indicated in figure 4.8: but if the compound has an electronic absorption band in the region under study, we obtain an anomalous O.R.D. curve, called a *Cotton effect* curve (figure 4.8). For example, if we measure the O.R.D. curve for an optically active ketone in the region of the n $\rightarrow$ π^* transition band (≈ 280 nm), we shall see an anomalous curve. As shown in figure 4.8, the Cotton effect can be positive or negative, depending on whether a peak or a trough is met first in going from long to short λ.

The shape of the Cotton-effect curves (positive or negative) can be correlated empirically with very many features of the stereochemistry around the chromophore, and while most work has centred on the weak n $\rightarrow$ π^* chromophore of carbonyl compounds, the Cotton effect can be extended to the study of any optically active molecule that possesses a chromophore.

The $\pi \rightarrow \pi^*$ transitions (of conjugated ketones, etc.) are of much higher intensity than n $\rightarrow$ π^* transitions and are consequently more difficult to study, and the multiple Cotton effect curves observed make interpretation more complex.

Anomalous curves in O.R.D. have their counterpart in circular dichroism curves (see figure 4.8): until recently, it was difficult to record the variation of C.D. with wavelength, especially near the chromophores that absorb in the vacuum ultraviolet, but new equipment renders this region accessible down to 165 nm.

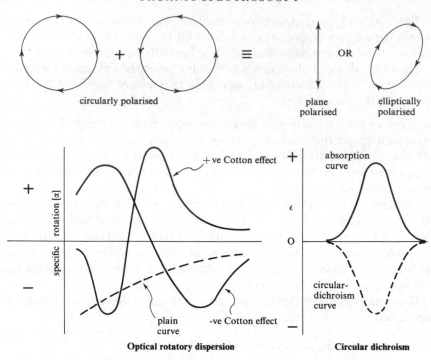

Figure 4.8 *Optical rotatory dispersion and circular dichroism curves.*

Chromophores successfully studied include aromatic systems, hetero-aromatic systems and, recently, groups that have absorptions only in the less accessible vacuum ultraviolet (such as OH groups in alcohols and carbohydrates).

One example of C.D. has even been recorded in the infrared region.

4S.5.3 The octant rule

The sign of the Cotton effect in O.R.D. and C.D. curves can be computed empirically from a knowledge of the absolute stereochemistry of substituents around the chromophore: conversely, the absolute stereochemistry of substituents can be deduced from the sign of the Cotton effect. The enormous potential in this has been fully exploited in a wide range of molecules but we can illustrate the principle of the method by reference to the *ketone octant rule*.

If an organic molecule is placed at the origin of three-dimensional ortho-gonal axes, the three orthogonal planes will cut the molecule into eight parts, each lying in an 'octant' defined by the intersecting co-ordinate planes (see figure 4.9). For cyclohexanones the molecule is placed as shown, with the $C{=}O$ group along the z axis, the xy plane bisecting the $C{=}O$ bond and the yz plane bisecting C_1 and C_4. The molecule is now viewed in projection

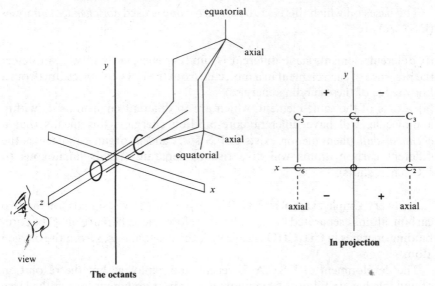

Figure 4.9   *The octant rule for ketones.*

along the z-axis, and except in rare cases the molecule can be seen to occupy only the four rear octants.

The octant rule, applied to cyclohexanones, states that (i) substituents lying *on the co-ordinate planes* have no influence on O.R.D., (ii) substituents lying in the *far upper left* and *far lower right* octants make a positive contribution to the sign of the Cotton effect: the other substituents make a negative contribution (see figure 4.9).

3-Methylcyclohexanone is an optically active ketone, and therefore can be treated by the octant rule. Write out the possible conformations of the dextro and laevorotatory forms (four forms in all) and decide which will exhibit positive or negative Cotton effects. Given that D-(+)-3-methylcyclohexanone exhibits a positive Cotton effect, what is the preferred conformation of this isomer?

4S.6   ELECTRON SPECTROSCOPY FOR CHEMICAL ANALYSIS—E.S.C.A.
Electronic-absorption spectra in the ultraviolet–visible region arise when electrons in n and π orbitals undergo upward transitions on bombardment by light: at higher energies (in the vacuum ultraviolet) transitions from σ orbitals are also observed. At still higher energies (using X-rays for bombardment) we can induce the nonvalency *core electrons* of the atoms to undergo transitions: thus if we bombard an organic molecule with $MgK\alpha_{1,2}$ or $AlK\alpha_{1,2}$ radiation, electrons will be ejected from the carbon 1s orbitals, and the spectrum of energies in these ejected electrons can be recorded as a direct measure of the *core binding energies* for C 1s orbitals.

The bases on which this *electron spectroscopy* is used *for chemical analyses* (E.S.C.A.) are

(i) different elements have different 1s binding energies, and we can detect the presence of each element in a molecule from the E.S.C.A. spectrum (from a knowledge of these binding energies).
(ii) atoms of the same element, which are in different environments within a molecule, will have different core-binding energies: this means that a *chemical shift* phenomenon exists in E.S.C.A., and in organic molecules the different carbon atoms will give rise to separate signals (analogous to ^{13}C n.m.r. shifts).

As a very simple example, CH_3CHO gives rise to two signals for the two carbon atoms, separated by 2.6 eV: therefore the difference in C 1s core-binding energies in CH_3CHO is 2.6 eV. A third signal arises from the oxygen atom.

The development of E.S.C.A. is recent and explosive but the resolution obtainable has up till now been poor: not only are inherent linewidths large (of the order 1.0 eV) but the X-radiation has been incoherent. X-ray mono-chromators are now coming into routine use and will vastly enhance the applicability of the technique because of the improved resolution obtainable. The method is complementary to n.m.r. (and other spectroscopic techniques) and has the advantage of high sensitivity: successful spectra have been recorded on mg quantities, and the method works equally well on solid, liquid or gas samples.

As in n.m.r. it is usually more important to an organic chemist to measure relative shifts than to measure the absolute core-binding energies, but these latter can be measured and correlated with theoretical values. Typical C 1s binding energies are around 290 eV: within hydrocarbons the chemical shifts are small, while in carbonium ions major shifts can be observed. Shifts have been correlated with electronegativity of substituent, it being found that increasing electronegativity (C, Br, N, Cl, etc.) leads to increases in the core-binding energies. It becomes possible therefore to deduce structural information from E.S.C.A. spectra as from n.m.r. spectra.

The time scale of E.S.C.A. is very short: the life-times of 'hole states' (those from which the core electron, usually 1s, has been ejected) are of the order 10^{-14}–10^{-17} s. The method is therefore well suited to studying fast reactions, transient species, radicals, carbonium ions, etc.: in radical studies E.S.C.A. gives similar information to e.s.r. (see section 3S.6).

Dramatic insights into carbonium ion structures have been achieved. We can now quantitatively compare the delocalisation of charge in CH_3CO^+ and $PhCO^+$, by measuring the E.S.C.A. shift between the CO carbon atom and the other carbon atoms of the ion structure. For CH_3CO^+ the shift is 6.0 eV while for $PhCO^+$ it is 5.1 eV: the lesser shift in $PhCO^+$ implies a lesser difference in core-binding energies and *therefore* greater delocalisation of charge.

In the *tert*-butyl cation the charge is substantially localised on C_1 (E.S.C.A. shift between C_1 and the methyl carbons is 3.4 eV), while in the trityl and tropylium cations only a single C 1s line is observed, showing that the charge is efficiently delocalised over all carbons in these systems.

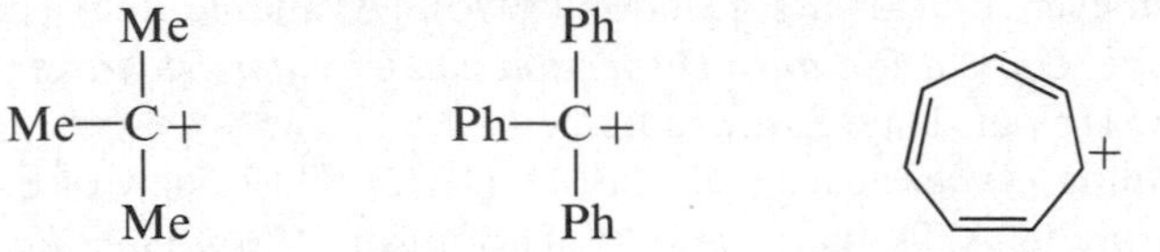

Me|Me—C+|Me	Ph|Ph—C+|Ph	
tert-butyl	trityl	tropylium

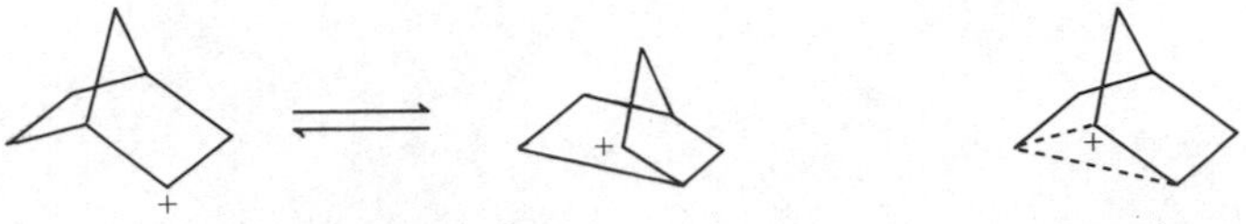

fast equilibrium   or   resonance
classical norbornyl   or   **nonclassical norbornyl**

In the norbornyl cation, the argument concerning its classical or non-classical nature hinges on whether we have a fast-equilibrating system or a true resonance system. By comparing the E.S.C.A. shifts within the molecules of a series of norbornyl cations Olah has shown that in the parent norbornyl cation the shift (1.7 eV) is much less than would be expected in a charge-localised equilibrating ion, and concludes that this ion is indeed nonclassical.

FURTHER READING

MAIN TEXTS

A. E. Gillam and E. S. Stern, *An Introduction to Electronic Absorption Spectroscopy in Organic Chemistry*, Arnold London (2nd ed., 1957).

R. A. Friedel and M. Orchin, *Ultraviolet Spectra of Aromatic Compounds*, Wiley, New York (1951). (These two books, although now rather old, contain a fund of data, spectra and interpretation.)

A. I. Scott, *Interpretation of the Ultraviolet Spectra of Natural Products*, Pergamon, Oxford (1964). (The authoritative work on Woodward rules, etc.)

SPECTRA CATALOGUES

Friedel and Orchin (1951) includes large numbers of spectra.

E. Clar, *Polycyclic Hydrocarbons*, Vols 1 and 2, Springer, Berlin and Academic Press, London (1965). (Very many condensed aromatic hydrocarbon systems described, with electronic spectra.)

DMS UV Atlas of Organic Compounds, Verlag Chemie, Weinheim and Butterworths, London. (Up-to-date series of well-presented spectra.)

Sadtler Ultraviolet Spectra, Sadtler, Pennsylvania, and Heyden, London.

SUPPLEMENTARY TEXTS

R. J. Taylor, *Fluorescence*, Unilever Educational Booklets: Advanced Series, Unilever, Blackfriars, London (1970). (Available free).

P. Crabbé, *Optical Rotatory Dispersion and Circular Dichroism in Organic Chemistry*, Holden-Day, San Francisco (1965).

C. Nordling, *Angew. Chem. int. Ed.*, **11** (1972), 83 (Review of E.S.C.A.).

C. R. Brundle, A. D. Baker, and M. Thompson, *Chem. Soc. Rev.*, **1** (1972), 355 (Review of E.S.C.A.).

5

MASS SPECTROSCOPY

As with n.m.r., the application of mass spectroscopy to organic chemistry has undergone explosive growth since around 1960.

Organic chemists use mass spectroscopy in two principal ways: (a) to measure *relative molecular masses* (molecular weights) with very high accuracy; from these can be deduced exact molecular formulae: (b) to detect within a molecule the places at which it *prefers to fragment*; from this can be deduced the presence of recognisable groupings within the molecule.

5.1 BASIC PRINCIPLES

In the simplest mass spectrometer (figure 5.1), organic molecules are bombarded with electrons and converted to highly energetic positively charged ions (*molecular ions*, or *parent ions*), which can break up into smaller ions (*fragment ions*, or *daughter ions*); the loss of an electron from a molecule leads to a radical cation, and we can represent this process as $M \rightarrow M^{+}$. The molecular ion M^{+} commonly decomposes to a pair of fragments, which may be either a radical plus an ion, or a small molecule plus a radical cation. Thus

$$M^{+} \rightarrow m_1^{+} + m_2 \cdot \quad \text{or} \quad m_1^{+} + m_2$$

The molecular ions, the fragment ions and the fragment radical ions are separated by deflection in a variable magnetic field according to their mass and charge, and generate a current (the *ion current*) at the collector in proportion to their *relative abundance*. A *mass spectrum* is a plot of relative abundance against the ratio mass/charge (the *m/e* value). For singly charged ions, the lower the mass the more easily is the ion deflected in the magnetic field. Doubly charged ions are occasionally formed: these are deflected much more than singly charged ions of the same mass, and they appear in the mass spectrum at the same value as singly charged ions of twice the mass, since $2m/2e = m/e$.

Neutral particles produced in the fragmentation, whether uncharged molecules (m_2) or radicals ($m_2 \cdot$), cannot be detected directly in the mass spectrometer.

Figure 5.1 shows a block layout for a simple mass spectrometer. Since the ions must travel a considerable distance through the magnetic field to the collector, very low pressures ($\approx 10^{-6}$–10^{-7} mm Hg $\equiv 10^{-4}$ N m^{-2}) must be maintained by the use of diffusion pumps.

Figure 5.2 shows a simplified line-diagram representation of the mass spectrum of 2-methylpentane (C_6H_{14}). The most abundant ion has an m/e value of 43 (corresponding to $C_3H_7{}^+$) showing that the most favoured point of rupture occurs between C_x and C_y: this most abundant ion (the *base peak*) is given an arbitrary abundance of 100, and all other intensities are expressed as a percentage of this (*relative abundances*). The small peak at m/e 86 is obviously the molecular ion. The peaks at m/e 15, 29 and 71 correspond to $CH_3{}^+$, $C_2H_5{}^+$ and $C_5H_{11}{}^+$, respectively, etc.: the fragment ions arise from

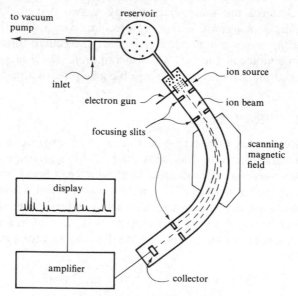

Figure 5.1 *A simple mass spectrometer. Molecules drift from the reservoir into the ion source, where they are ionised by electron bombardment. The resulting ion beam consists of molecular ions, fragment ions and neutral fragments; the ions are deflected by the magnetic field onto the collector.*

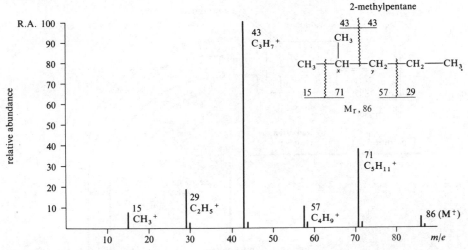

Figure 5.2 *Simplified mass spectrum of 2-methylpentane showing how the chain can break in several places; but breaks occur preferentially at the branching point, the most abundant ion being m/e 43 ($C_3H_7^+$).*

the rupture of the molecular ion, either directly or indirectly, and the analysis of many thousands of organic mass spectra has led to comprehensive semi-empirical rules about the preferred *fragmentation modes* of every kind of organic molecule. The application of these rules to organic structural elucidation will be developed later in the chapter.

The mass spectrum of a compound can be obtained on a smaller sample size (*in extremis* down to 10^{-12} g) than for any other of the main spectroscopic techniques, the principal disadvantages being the destructive nature of the process which precludes recovery of the sample, the difficulty of introducing small enough samples into the high-vacuum system needed to handle the ionic species involved and the high cost of the instruments. Mass spectroscopy is unlike the other spectroscopic techniques met in this book in that it does not measure the interaction of molecules with the spectrum of energies found in the electromagnetic spectrum, but the output from the instrument has all other spectroscopic characteristics, in showing an array of signals corresponding to a spectrum of energies: only the purist would unprofitably deny mass spectroscopy its name.

5.2 INSTRUMENTATION—THE MASS SPECTROMETER

We can conveniently study the design of a sophisticated mass spectrometer by considering each of its five main parts successively, with reference to figure 5.3.

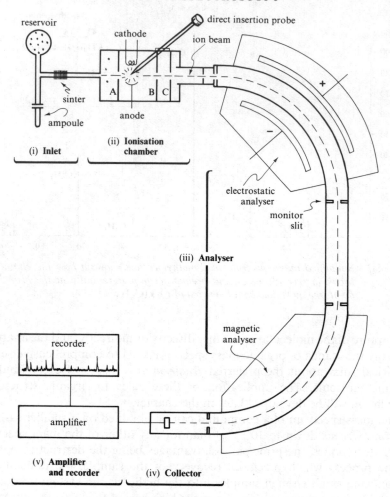

Figure 5.3   *High-resolution, double-focusing mass spectrometer. (Evacuation system not shown.)*

5.2.1   Sample Insertion—Inlet Systems

Organic compounds that have moderate vapour pressures at temperatures up to around 300° (including gases) can be placed in an ampoule connected via a reservoir to the ionisation chamber. Depending on volatility, it is possible to cool or heat the ampoule, etc. to control the rate at which the sample volatilises into the reservoir, from which it will diffuse slowly through the sinter into the ionisation chamber. Samples with lower vapour pressures (for example solids) are inserted directly into the ionisation chamber on the end of a probe, and their volatilisation controlled by heating the probe tip.

5.2.2   Ion Production in the Ionisation Chamber

The several methods available for inducing the ionisation of organic compounds are discussed in section 5S.1, but electron bombardment is routinely
used. Organic molecules react on electron bombardment in two ways:
either an electron is captured by the molecule, giving a radical anion, or an
electron is removed from the molecule giving a radical cation.

$$M \xrightarrow{\;e\;} M^{\overline{\cdot}} \quad \text{or} \quad M \xrightarrow{\;-e\;} M^{\overset{+}{\cdot}} + 2e$$

The latter is more probable by a factor of 10^2, and positive-ion mass spectroscopy is the result.

Most organic molecules form molecular ions $(M^{\overset{+}{\cdot}})$ when the energy of the
electron beam reaches 10–15 eV ($\approx 10^3$ kJ mol^{-1}). While this minimum
ionisation potential is of great theoretical importance, fragmentation of the
molecular ion only reaches substantial proportions at higher bombardment
energies, and 70 eV ($\approx 6 \times 10^3$ kJ mol^{-1}) is used for most organic work.

When the molecular ions have been generated in the ionisation chamber
they are expelled electrostatically by means of a low positive potential on
a repeller plate (A) in the chamber. Once out, they are accelerated down the
ion tube by the much higher potential between the accelerating plates B and C
(several thousand volts). Initial focusing of the ion beam is effected by a
series of slits.

5.2.3   Separation of the Ions in the Analyser

Theory. In a magnetic analyser ions are separated on the basis of m/e values,
and a number of equations can be brought to bear on the behaviour of ions
in the magnetic field.

The kinetic energy E of an ion of mass m travelling with velocity v is
given by the familiar $E = \frac{1}{2}mv^2$. The potential energy of an ion of charge e
being repelled by an electrostatic field of voltage V is eV. When the ion is
repelled, the potential energy eV is converted into the kinetic energy $\frac{1}{2}mv^2$,
so that

$$eV = \tfrac{1}{2}mv^2$$

therefore

$$v^2 = \frac{2eV}{m} \qquad\qquad (5.1)$$

When ions are shot into the magnetic field of the analyser they are drawn into
circular motion by the field, and at equilibrium the centrifugal force of the ion
$(mv^2 r)$ is equalled by the centripetal force exerted on it by the magnet (eBv),
where r is the radius of the circular motion and B is the field strength. Thus

$$\frac{mv^2}{r} = eBv$$

therefore

$$v = \frac{eBr}{m} \tag{5.2}$$

Combining equations (5.1) and (5.2)

$$\frac{2eV}{m} = \frac{eBr^2}{m}$$

therefore

$$\frac{m}{e} = \frac{B^2r^2}{2V} \tag{5.3}$$

It is from equation (5.3) that we can see the inability of a mass spectrometer to distinguish between an ion m^+ and an ion $2m^{2+}$, since their ratio m/e has the same value of $(B^2r^2/2V)$, and these three parameters B, r and V dictate the path of the ions.

To change the path of the ions so that they will focus on the collector and be recorded, we can either vary V (the accelerating voltage) or B (the strength of the focusing magnet). *Voltage scans* (the former) can be effected much more rapidly than *magnetic scans* and are used where fast scan speed is desirable.

Resolution. The ability of a mass spectrometer to separate two ions (the spectrometer resolution) is acceptably defined by measuring the depth of the valleys between the peaks produced by the ions. If two ions of m/e 999 and 1000, respectively, can just be resolved into two peaks such that the recorder trace almost reaches back down to the baseline between them, leaving a valley which is 10 per cent of the peak height, we say the resolution of the spectrometer is '1 part in 1000 (10 per cent valley resolution)'. Simple magnetic-focusing instruments have resolving powers of around 1 in 7500 on this basis.

Double focusing. Ions repelled by the accelerator plates do not in practice all have identical kinetic energies and this energy spread constitutes the principal limiting factor in improving the resolution of a magnetic analyser. Preliminary focusing can be carried out by passing the ion beam between two curved plates, which are electrostatically charged—an *electrostatic analyser*.

Different equations hold for the behaviour of ions in an electrostatic (rather than magnetic) field. Here, the centrifugal force $(mv^2/r) = eB$, and combining this with $eV = \frac{1}{2}mv^2$ gives

$$r = \frac{2V}{B} \tag{5.4}$$

Thus the radial path followed by ions of a given velocity is independent of m and e (and m/e values). The electrostatic analyser focuses ions of identical kinetic energy onto the monitor slit, whatever their m/e values, and coupled with a magnetic analyser to resolve m/e values this *double-focusing* spectrometer can attain resolutions of 1 in 60 000.

In a low-resolution instrument we may identify an ion of m/e 28 as possibly CO^+ or N_2^+ or $C_2H_4^+$; the accurate mass measurement possible at high resolution enables us to distinguish among several possible exact formula masses, for example $CO^+(27.9949)$, $N_2^+(28.0062)$, $C_2H_4^+(28.0312)$.

5.2.4 THE COLLECTOR
The small current produced at the collector plate (the *ion current*) is fed to the amplifier–recorder system, but details of the devices used to detect ion impact lie in the electronic engineer's domain.

5.2.5 THE AMPLIFIER–RECORDER
Two essential features of the recording system in a mass spectrometer are that it must (a) have a very fast response, and be able to scan several hundred peaks per second, and (b) be able to record peak intensities varying by a factor of more than 10^3. The problem of fast response can be met by using mirror galvanometers, which reflect an ultraviolet light beam onto fast-moving photographic paper. Problem (b) is commonly overcome by having a series of mirror galvanometers covering a range of sensitivities (for example three covering the ratio $1:10:100$, or five covering the ratio $1:3:10:30:100$). Simultaneous scanning produces the mass spectrum trace shown in figure 5.4(a). An alternative solution is to make the galvanometer response logarithmic, giving the trace shown in figure 5.4(b).

To measure the m/e values on a recorder trace manually, we have to count in units from reference peaks, there being in general an observable peak at every mass number. The relative abundances are measured without difficulty from the galvanometer sensitivity ratio, or using a logarithmic measuring device if the trace is presented as in figure 5.4(b).

The most directly informative presentation of mass spectroscopic information is in the form of a line diagram or bar diagram such as in figure 5.2 and this is easily constructed from the extracted data. Computerised printouts contain much information, but all of this need not be significant to the organic chemist's needs; often tabular presentation of (say) the 20 most abundant ions together with m/e values is sufficient.

5.3 ISOTOPE ABUNDANCES

Few elements are monoisotopic, and table 5.1 gives the natural isotope abundance of the elements we might expect to encounter in organic compounds. Ions containing different isotopes appear at different m/e values.

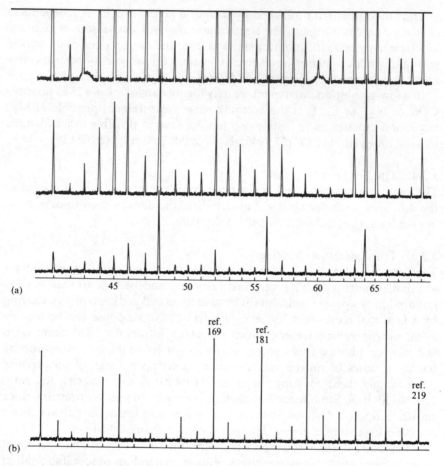

Figure 5.4 *Typical appearance of mass spectra* (a) *with three mirror galvanometers scanning simultaneously and* (b) *with a logarithmic galvanometer. Note the metastable ions at m/e 43.4 and m/e 60.2 in* (a).

For an ion containing *n* carbon atoms, there is a probability that approximately $1.1n$ per cent of these atoms will be ^{13}C, and this will give rise to an ion of mass one higher than the ion that contains only ^{12}C atoms. The molecular ion for 2-methylpentane (figure 5.2) has an associated M + 1 ion, whose intensity is approximately 6.6 per cent that of the molecular ion; while the molecular ion has only ^{12}C atoms the M + 1 ion contains ^{13}C atoms. (The contribution of ^2H atoms should not be overlooked, even though the probability of their presence is not as high as for ^{13}C, and in exact work the statistics of both ^{13}C and ^2H abundances must be calculated.)

A second associated peak can arise at *m/e* M + 2 if two ^{13}C atoms are present in the same ion (or if two ^2H atoms, or one ^{13}C and one ^2H are present); these probabilities can be calculated and may be a help in deciding

Table 5.1 Natural isotope abundances and relative atomic masses ($^{12}C = 12.000000$)
for common elements

Isotope	Natural abundance (per cent)	Relative atomic mass
^1H	99.985	1.007825
^2H	0.015	2.014102
^{12}C	98.9	12.000000
^{13}C	1.1	13.003354
^{14}N	99.64	14.003074
^{15}N	0.36	15.000108
^{16}O	99.8	15.994915
^{17}O	0.04	16.999133
^{18}O	0.2	17.999160
^{19}F	100	18.998405
^{28}Si	92.2	27.976927
^{29}Si	4.7	28.976491
^{30}Si	3.1	29.973761
^{31}P	100	30.973763
^{32}S	95.0	31.972074
^{33}S	0.76	32.971461
^{34}S	4.2	33.967865
^{35}Cl	75.8	34.968855
^{37}Cl	24.2	36.965896
^{79}Br	50.5	78.918348
^{81}Br	49.5	80.916344
^{127}I	100	126.904352

the formula for an ion in the absence of exact mass measurement. For example the two ions $C_8H_{12}N_3{}^+$ and $C_9H_{10}O_2{}^+$ have the same unit mass (m/e 150), and the M + 1 relative abundances are similar (9.98 and 9.96 per cent respectively); the M + 2 abundances are, however, sufficiently different to enable differentiation of the structures (0.45 and 0.84 per cent, respectively). The ability to see M + 2 peaks of such low abundance depends on there being a large M^{+} peak.

Ions containing one bromine atom create a dramatic effect in the mass spectrum because of the almost equal abundance of the two isotopes; pairs of peaks of roughly equal intensity appear, separated by two mass units.

Equally characteristic are the ions from chlorine compounds engendered by the ^{35}Cl and ^{37}Cl isotopes; for ions containing one Cl atom, the relative intensities of the lines, separated by two mass units, is 3:1. Ions containing one sulphur atom also have associated m + 2 peaks.

The picture becomes much more complicated when one considers the relative abundances of ions containing several polyisotopic elements; the presence of two bromine atoms in an ion gives rise to three peaks at m, m + 2, and m + 4, the relative intensities being 1:2:1, while for three bromines the peaks arise at m, m + 2, m + 4, m + 6, with relative intensities 1:3:3:1. These figures ignore any contribution from ^{13}C that may be present.

For each element in a given ion, the relative contributions to m + 1, m + 2 peaks, etc. can be calculated from the binomial expansion of $(a + b)^n$, where a and b are the relative abundances of the isotopes, and n the number of these atoms present in the ion. Thus for three chlorine atoms in an ion, expansion gives $a^3 + 3a^2b + 3ab^2 + b^3$. Four peaks arise; the first contains three ^{35}Cl atoms and each successive peak has ^{35}Cl replaced by ^{37}Cl until the last peak contains three ^{37}Cl atoms. The m/e values are separated by two mass units, at m, m + 2, m + 4, m + 6. Since the relative abundances of ^{35}Cl and ^{37}Cl are 3:1 (that is $a = 3$, $b = 1$) the intensities of the four peaks (ignoring contributions from other elements) are $a^3 = 27$, $3a^2b = 27$, $3ab^2 = 9$, $b^3 = 1$ (that is 27:27:9:1).

5.4 THE MOLECULAR ION

5.4.1 STRUCTURE OF THE MOLECULAR ION

In previous sections we represented the molecular ion as M^+, signifying a radical cation produced when a neutral molecule loses an electron: where does the electron come from?

For electron bombardment around their ionisation potentials (10–15 eV $\approx 10^3$ kJ mol^{-1}) it is meaningfully possible in organic molecules to say which are the likeliest orbitals to lose an electron. The highest occupied orbitals of aromatic systems and nonbonding orbitals on oxygen and nitrogen atoms readily lose an electron; the π-electrons of double and triple bonds are also vulnerable. At the instant of ionisation in a Franck–Condon process, before any structural rearrangement can occur, these ionisations can then be represented as shown below. In alkanes, all we can say is that the ionisation of C—C σ-bonds is easier than that of C—H bonds.

In the strictest sense, however, we must not attempt to localise the excitation produced by the loss of an electron when the electrons themselves are not localised: in the absence of specific evidence we must be satisfied with the generalised view of an electron being expelled from the whole molecular orbital and of the excitation energy being spread throughout the molecule. For electron bombardment of complex molecules around 70 eV (6×10^3 kJ mol^{-1}) any specificity in the site of electron removal is lost entirely. These arguments apply also to the structures of fragment ions and fragment radical ions, and these are frequently written in square brackets, for example $[C_5H_5]^+$ or $[C_4H_7]^+$, no attempt being made to speculate on the precise structures. The use of a partial square bracket (for example $C_5H_5]^{\ddagger}$ or $C_4H_7]^{\ddagger}$, or as in structure I below) is a useful alternative, especially for larger structures. Organic chemists have, however, successfully adapted the postulates of resonance theory to explain the reactivities of functional groups and the mechanistic processes inherent in syntheses, rearrangements and degradations. It is not surprising that they have extended their mechanistic interpretations to include fragmentation reactions of molecular ions, and in so doing they have perhaps oversimplified the problems faced by theoretical workers. The organic chemist's argument is that it may be inexact to write the single structure II rather than I for the molecular ion of 2-propanol after 70 eV bombardment: but II makes it easy to *rationalise* its fragmentation to a methyl radical (CH$_3$ ·) and the ion $[C_2H_5O]^+$ at m/e 45.

$$\text{I} \qquad\qquad \text{II} \qquad\qquad \longrightarrow \text{CH}_3\cdot \qquad m/e\ 45$$

At the instant of fragmentation, it is certainly true that sufficient excitation energy must be concentrated within the appropriate σ-bond to exceed the energy of activation for its rupture. Provided mechanistic notations are used with full acknowledgement of their implications and limitations, they can supply to organic chemists a familiar rational framework for the interpretation of fragmentation processes that cannot be supplied by the unadulterated precision of molecular-orbital theory.

The fragmentation of molecular ions, etc., is discussed systematically in section 5.6.

5.4.2   RECOGNITION OF THE MOLECULAR ION

The molecular ions of roughly 20 per cent of organic compounds decompose so rapidly ($< 10^{-5}$ s) that they may be very weak or undetected in a routine 70 eV spectrum. For most unknown compounds the ion cluster appearing at highest m/e value is likely to represent the molecular ion with its attendant M + 1 peaks, etc., but we must apply a number of tests to ensure that this is so.

Abundant molecular ions are given by aryl amines, nitriles, fluorides and chlorides. Aromatic hydrocarbons and heteroaromatic compounds give strong M$^+$ peaks provided no side-chain of C$_2$ or longer is present: indeed the $M^{\ddagger}$ peak is often the base peak, and doubly charged ions may often be observed in the mass spectra of these compounds appearing at $m/2e$ values. Aryl bromides and iodides lose halogen too readily to give strong molecular ion peaks. Other classes with weakened $M^{\ddagger}$ peaks are aryl ketones (which fragment easily to $ArCO^+$) and benzyl compounds such as side-chain hydrocarbons ($ArCH_2R$) or ($ArCH_2X$) (both of which fragment at the benzylic carbon).

Absence of molecular ions (or an extremely weak $M^{\ddagger}$ peak) is characteristic of highly branched molecules whatever the functional class. Alcohols and molecules with long alkyl chains also fragment easily and lead to very weak $M^{\ddagger}$ peaks.

Isotope abundances should correlate with the appearance of the purported molecular-ion cluster as discussed in section 5.3. The intensities of $M + 1$, $M + 2$ peaks, etc., are obviously most easily measured and of greatest value when the $M^{\ddagger}$ peak is fairly abundant.

Nitrogen-containing compounds with an *odd* number of nitrogen atoms in the molecule must have an *odd* molecular weight (relative molecular mass).

An *even* number of nitrogen atoms, or no nitrogen at all, leads to an *even* molecular weight (relative molecular mass).

Common fragment ions in the spectrum contribute positive support for the assignment of $M^{\ddagger}$; what constitutes 'common fragment ions' is the subject of section 5.6.

Unusual fragment ions should make one suspicious: for example, molecular ions can give rise to a series of weaker ions at $M - 1$, $M - 2$ and $M - 3$ due to successive loss of hydrogen, but a *specific* fragmentation leading directly to $M - 3$ (or $M - 4$ or $M - 5$) is never observed. An apparent $M - 14$ peak (corresponding to direct loss of a CH_2 fragment) is so rare that the purported $M^{\ddagger}$ ion should be discounted; a more likely explanation is the presence of a lower homologue as contaminant in the sample.

5.4.3 MOLECULAR FORMULA FROM THE MOLECULAR ION

For an unknown organic compound, the ability to measure its relative molecular mass (molecular weight) to within four decimal places leads immediately to an accurate molecular formula for the compound.

Accurate relative atomic masses (atomic weights) for the principal isotopes of the elements commonly met in organic chemistry are given in table 5.1 on page 193.

As an example, consider an unknown compound X whose relative molecular mass (to the nearest integer) is measured at low resolution to be 100. From this and other evidence the compound could be either A $C_6H_{12}O$ or B $C_4H_4O_3$. High-resolution mass measurement of the molecular ion gives m/e 100.08871, proving that the correct structure is A.

Mass measurement can be carried out at its most sophisticated level by having a high-resolution instrument interfaced to a computer, which is programmed to identify reference peaks and to interpolate from their known masses to the masses of other individual ions from the sample. An additional program can convert the accurate masses to the element compositions of the ions, so that the print-out contains a list of ion masses, abundances and compositions. The positions of known accurate masses can be obtained from an electronic *mass marker* or by using reference compounds such as per-fluorokerosene, PFK (mixed long-chain perfluoroalkanes, C_nF_{2n+2}): PFK fragments to a series of well-spaced fragment ions up to high m/e values and in figure 5.4(b) the reference peaks are from PFK. In *double-beam* spectrometers PFK can be fed simultaneously into the instrument with the sample, and the ion masses of the sample displayed in parallel with those of the PFK.

An alternative method of mass measurement is to couple the output of the instrument to a cathode-ray oscilloscope, and to display alternately on the oscilloscope tube the peak whose mass is to be measured and an ion of known mass from the reference compound. The accelerator voltage of the spectro-meter can then be adjusted until these two peaks overlap, and the difference in mass between them can be calculated as a function of the change in acceler-ator voltage. This process is known as *peak matching*.

5.5 METASTABLE IONS

5.5.1 THE NATURE OF METASTABLE IONS

In the diagrammatic mass spectrum in figure 5.4(a) can be seen broad peaks at noninteger masses m/e 60.2 and m/e 43.4. The ions producing these peaks are termed *metastable ions*: they have lower kinetic energy than normal ions, and arise from fragmentations that take place during the flight down the ion tube rather than in the ionisation chamber. The exact position where they are formed in the tube determines whether or not we can easily observe them.

Up until now we have tacitly assumed that molecular ions formed in the ionisation chamber do one of two things: either they decompose completely and very rapidly in the ion source and never reach the collector (as in the case of highly branched molecular ions with lifetimes less than $\approx 10^{-5}$ s), or else they survive long enough to reach the collector and be recorded there (life-times longer than $\approx 10^{-5}$ s). We have also assumed that fragment ions are produced by decompositions of a proportion of the molecular ions in the ion source.

Depending on the inherent stability of an ion, and on the amount of ex-citation energy absorbed on bombardment, ion lifetimes will vary in a com-plex manner: a given molecular ion may possess a spread of energies, and not surprisingly, some of the molecular ions will have intermediate lifetimes ($\approx 10^{-5}$ s) and so leave the ionisation chamber intact, but decompose *en route* to the collector. Some fragment ions behave similarly, decomposing between ion source and collector.

Suppose that a large number of molecules of M are converted to molecular ions $M^{\ddagger}$: not all of the $M^{\ddagger}$ ions will possess the same excitation energy and therefore some will have longer lifetimes than others.

The $M^{\ddagger}$ ions with shortest lifetimes may decompose in the ionisation chamber to stable daughter ions A^+ and radicals $B\cdot$; the daughter ions A^+ will be detected at the collector normally. The molecular ions that leave the ion source intact will be accelerated by the accelerator voltage and will then possess a translational energy eV. Some of these $M^{\ddagger}$ ions may survive intact to the collector and be detected normally. If, however, others of these $M^{\ddagger}$ ions decompose to A^+ and $B\cdot$ immediately *after* acceleration, the translational energy of the parent $M^{\ddagger}$ (eV) will be shared *between* A^+ and $B\cdot$ in proportion to their masses (principle of conservation of momentum).

The translational energy of the daughter ion A^+ must then be *lower* than that of the parent ion, and this ion A^+ will arrive at the collector differently from the 'normal' A^+ ion produced in the ion source.

The ion A^+ with 'abnormal' translational energy is a *metastable ion*.

Note that metastable A^+ ions *have the same mass* as normal A^+ ions, but simply have less translational energy.

5.5.2   Ion Tube Regions

Of the metastable ions produced in the ion tube, only a fraction come to reasonable focus at the collector (unless by the use of special techniques): we must consider the successive regions of the ion tube to understand why this should be so.

The first field-free region, in a double-focusing instrument, lies between the ion source and the electrostatic analyser. (This region has no counterpart in single-focusing instruments.) If a metastable ion is produced here, it will be focused out by the electrostatic analyser because of its abnormal kinetic energy. Such ions, and any metastable ions formed *in the analyser* will appear out of focus (randomly) at the collector as background current, and will be undetected.

The second field-free region in a double-focusing instrument lies between the electrostatic and magnetic analysers. (In single-focusing spectrometers the corresponding region is between the ion source and the magnetic analyser.) Metastables produced in this region will be focused reasonably sharply by the magnetic analyser on the bases of their masses and translational energies, but since a metastable A^+ ion has the same mass and lower translational energy (less momentum) than the normal A^+ ion, the metastable A^+ ion is deflected more easily in the analyser than the normal A^+ ion: these metastable A^+ ions will appear on the spectrum among ions of lower mass, and are the only metastable ions detected routinely. *Metastable peaks* are broadened for a number of reasons, one of which is the possibility that some of the excitation energy leading to bond rupture may be converted to additional kinetic energy.

Ions produced *in the magnetic analyser* will be focused at the collector, but there will be substantial differences in energy between those formed at the *beginning* compared to those formed at the *end* of the analyser: this produces a continuum of low-intensity signals between the positions of normal A^+ and metastable A^+, and is usually too weak to be detected.

The third field-free region lies between the magnetic analyser and the collector. Since no focusing takes place in this region, a parent ion is already immutably on path, and if it decomposes to the metastable A^+ then this metastable will continue on the same path as the parent ion. The metastable ion is detected at the same m/e value as the parent ion.

5.5.3   CALCULATION OF METASTABLE ION m/e VALUES

The apparent mass of a metastable ion A^+ (m^*) can be calculated fairly accurately from the masses of the parent ion (m_1) and the normal daughter ion A^+ (m_2) from the equation

$$m^* = \frac{(m_2)^2}{m_1}$$

This equation often gives an apparent mass 0.1 to 0.4 mass units lower than is in fact observed.

As an example, the mass spectrum of toluene shows strong peaks at m/e 91 and m/e 65, together with a strong broad metastable peak at m/e 46.4. Now $65^2/91 = 46.4$, so we can interpret that the ion m/e 91 is decomposing by loss of 26 mass units to the daughter ion m/e 65, and that some of this fragmentation takes place in the second field-free region, leading to a metastable ion peak of m/e 46.4. (The value 46.4 is the *apparent* mass of the metastable ion, the *real* mass being the same as that of the normal daughter, that is, 65.)

It is worth restating that some of the m/e 91 ions will decompose before and after the second field-free region, producing m/e 65 ions, which are *not* focused at m/e 46.4.

Tables and computer programs are widely used to relate metastable ions to the corresponding parent and daughter ions; the use of the nomogram in figure 5.5 for this purpose is self-explanatory.

5.5.4   SIGNIFICANCE OF METASTABLE IONS

The presence of a metastable ion in a mass spectrum is taken as very good evidence that the parent ion undergoes decomposition *in one step* to the daughter, so that it is of considerable mechanistic importance to investigate metastable ions.

It follows from the discussion in section 5.5.2 that there may be one-step processes occurring in the mass spectrometer that do not produce metastable-ion peaks, so the absence of such peaks cannot be used to infer the absence of a one-step transition.

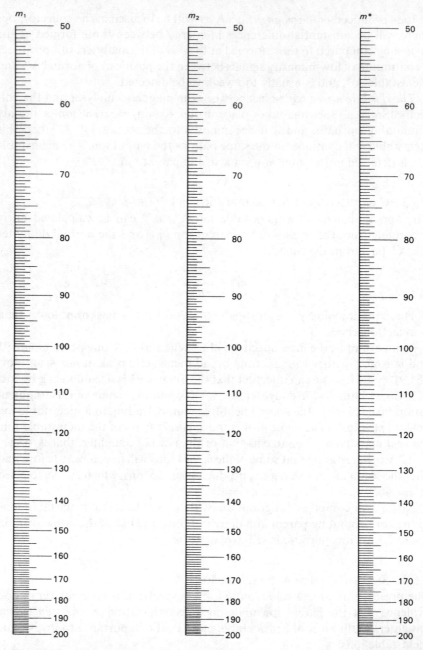

Figure 5.5 *Nomogram to aid identification of parent and normal daughter ions from the position of a related metastable ion. (Reproduced with permission from J. H. Beynon, Mass Spectrometry and its Application to Organic Chemistry, Elsevier, Amsterdam (1960).)*

5.6 FRAGMENTATION PROCESSES

5.6.1 REPRESENTATION OF FRAGMENTATION PROCESSES

If a molecular ion loses a methyl radical ($CH_3 \cdot$) the mass spectrum will show an ion 15 mass units below the molecular ion: we can write this process as

$$M^{\ddagger} \rightarrow CH_3^{\cdot} + (M - 15)^{+}$$

An alternative is to write the fragment ion as $M - CH_3$. Throughout this chapter we shall use this convention frequently, referring to ions as $M - 18$, $M - 24$, $M - CO$, $M - H_2S$, etc., it being understood that these may in fact be even-electron ions, for example $(M - 15)^{+}$, or odd-electron radical ions, for example $(M - 18)^{\ddagger}$.

The same convention can be used to represent fragment ions as m^{+} (or $m^{\ddagger}$), and these may in turn fragment to $m - 1$, $m - CH_3$, $m - C_2H_5$, etc.

We saw in section 5.4 that it is on the whole not possible to write the structures of molecular ions with any degree of certainty, but in the absence of structural proof it is a legitimate convenience to indicate that ionisation takes place from the most easily ionised orbitals in the molecule.

In representing the molecular ion of an alcohol as I, we at least give ourselves a framework for discussing its behaviour on fragmentation. For more complex molecules such as II, we may find it expedient to present one decomposition as proceeding from the molecular ion III, and another as proceeding from IV or V. In doing so, we are using an extension of resonance theory.

R—Ö—H

      I   II   III   IV   V

These are of course postulates for ion structures; some techniques which help to verify the postulates are discussed in section 5S.3.

When we come to represent on paper the mechanics of how a particular fragment ion (m^{+}) is produced from a molecular ion ($M^{\ddagger}$), the difficulties are twice compounded since we may have no precise knowledge of the molecular-ion structure, and we cannot always know with precision the structure of the fragment ion (although we may know its element composition).

We shall proceed to write such mechanisms, largely because most of them are similar in type to the more familiar mechanisms of 'wet' organic chemistry, and are therefore a help in classifying the observable facts concerning the fragmentations associated with functional groups. To do so we must make an initial assumption that ions, during fragmentation, undergo minimum structural change. This is often contrary to observations carried out on the randomisation of atoms (particularly H) during fragmentation, and notable examples will be highlighted: isotope labelling is here an enormously important adjunct, but its discussion will be delayed until section 5S.3.

5.6.2 BASIC FRAGMENTATION TYPES AND RULES

Electron-pair processes (such as bond heterolysis) are represented by the conventional curved arrow, and one-electron processes (such as bond homolysis) by the fish-hook arrow: many fragmentations can either be represented as occurring by one-electron or as two-electron processes and mechanisms given in this book do not claim to represent actuality. The molecular ion, because of its excess energy, may take part in processes that have no counterpart in test-tube chemistry.

σ-*Bond rupture in alkane groups.* This can really only be represented by assuming that, at the instant of ionisation, sufficient excitation energy is concentrated on the rupturing bond to ionise it.

$$RCH_2-CH_2R' \equiv RCH_2:CH_2R' \xrightarrow{-e} RCH_2^{+}CH_2R'$$

$$\downarrow$$

$$RCH_2^{+} + \cdot CH_2R'$$

similarly $R-H \equiv R:H \xrightarrow{-e} R^{+}H \longrightarrow R^{+} + H\cdot$

σ-*Bond rupture near functional groups.* This may be facilitated by the easier ionisation of that group's orbitals, as in alcohols, where the nonbonding orbitals of oxygen are more easily ionised than the σ-orbitals.

Other groups in this category are ethers, carbonyl groups and compounds containing halogen, nitrogen, double bonds, phenyl groups, etc.

Elimination by multiple σ-bond rupture. Elimination by multiple σ-bond rupture may occur, leading to the extrusion of a neutral molecule such as CO,

(a)

one-electron mechanism

alternatively

one-electron mechanism

(b) 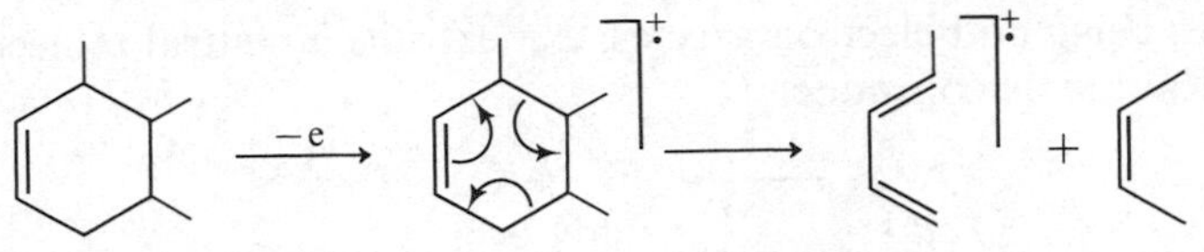 $\xrightarrow{-e}$ ... $\longrightarrow$ ... + ... two-electron mechanism

(c) 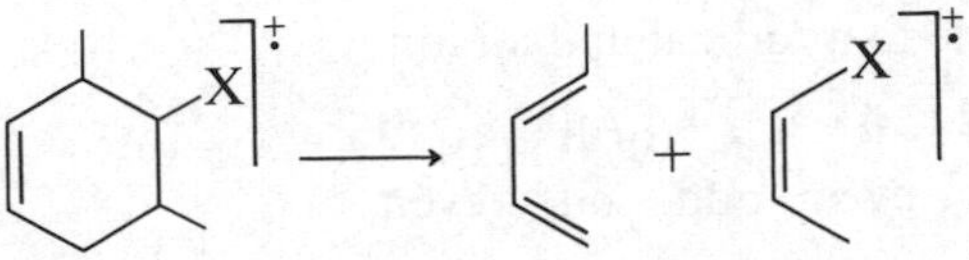 ...X... $\longrightarrow$ ... + ...X...

C_2H_4, C_2H_2, etc. A well-known example is the retro-Diels–Alder reaction of cyclohexenes, which can be represented as in (a) or (b). Highly stabilised ene fragments may cause charge retention to be in part reversed as in (c).

Rearrangements. These are common, the most frequently encountered having been described by F. W. McLafferty (see McLafferty's book, page 123), and named after him. It is exemplified in the case of a carbonyl compound (I) by the extrusion of an alkene, but is also exhibited by ions such as II, III, IV, etc.

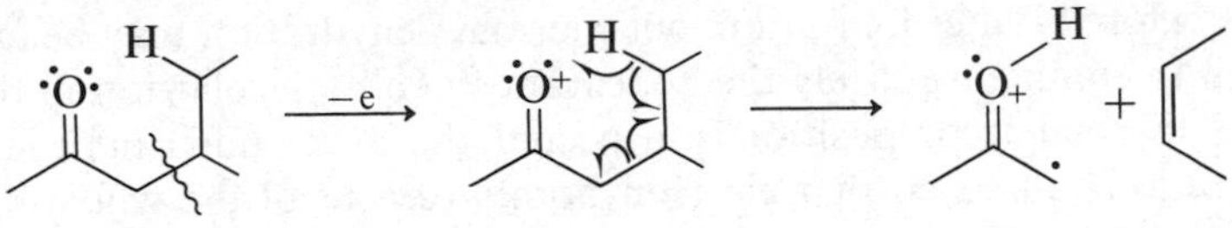

I $\xrightarrow{-e}$ ... $\longrightarrow$ ... + ...

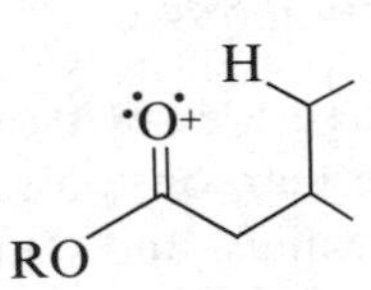

RO — ester — II

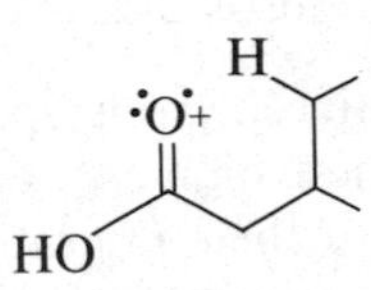

HO — carboxylic acid — III

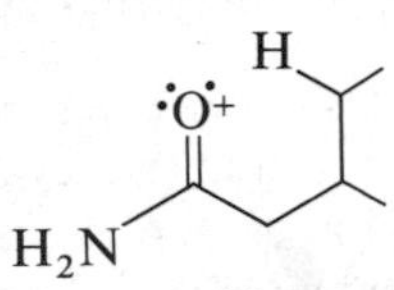

H_2N — amide — IV

The even-electron rule. The even electron rule is a rule-of-thumb interpretation of sound thermodynamic principles: in essence it states that an even-electron species (an ion, as opposed to a radical ion) will not normally fragment to two odd-electron species (that is, it will not degrade to a radical and a radical ion), since the total energy of this product mixture would be too high

$$A^+ \xcancel{\longrightarrow} B^+ + C^{\cdot}$$
even   odd   odd

In preference, an ion will degrade to another ion and a neutral molecule

$$A^+ \longrightarrow B^+ + C$$
even   even   even

Radical ions, being odd-electron species, can extrude a neutral molecule, leaving a radical ion as coproduct

$$A^{\ddagger} \longrightarrow B^{\ddagger} + C$$
$$\text{odd} \qquad\qquad \text{odd} \quad \text{even}$$

Radical ions can also degrade to a radical and an ion

$$A^{\ddagger} \longrightarrow B^{+} + C^{\cdot} (\text{or } B^{\cdot} + C^{+})$$
$$\text{odd} \qquad\qquad \text{even} \quad \text{odd} \quad \text{odd} \quad \text{even}$$

5.6.3 FACTORS INFLUENCING FRAGMENTATIONS

Functional groups. Some functional groups may direct the course of fragmentation profoundly, while other functional groups may have little effect. This is discussed fully in section 5.7.

Thermal decomposition. Thermal decomposition of thermolabile compounds may occur in the ion source, and commonly leads to difficulty in interpreting the mass spectra of alcohols, which may dehydrate *before* ionisation. In the case of alcohols, loss of water gives rise to a peak at M − 18 whether the loss occurs before or after ionisation, but thermal dehydration may be extensive enough to eliminate entirely the appearance of a molecular ion in the spectrum. If thermal decomposition is suspected, the compound can be ionised in a *cooled* ion source, so that electron bombardment of the whole molecule takes place. An alternative solution is to convert the alcohol to the more volatile trimethylsilyl derivative; this is discussed in section 5S.4.

Bombardment energies. For routine organic spectra these are ≈ 70 eV. It is worth noting that even with these high energies, molecular ions possess a maximum of ≈ 6 eV in excess of their ionisation potentials, and there is little change in the fragmentation *pattern* if this 70 eV is reduced to ≈ 20 eV; the *ion yield* (that is, the efficiency of ionisation) is, however, reduced, and the spectra are weaker in intensity overall. From ≈ 20 eV down to the ionisation potential of the molecule the spectrum becomes progressively simpler, since only the most favoured fragmentations are occurring: recording low-energy spectra is therefore a useful tool in bond-energy studies. It follows from these observations that the relative abundances of ions in a spectrum are only reproducible when bombardment energies are constant.

Relative rates of competing fragmentation routes. These also are important in dictating relative abundances. In the simple case of $A^{\ddagger}$ going either to B^{+} and $C^{\cdot}$ or to $B^{\ddagger}$ and C, the equilibrium abundances of $A^{\ddagger}$, B^{+} and $B^{\ddagger}$ depend on the relative rate constants for the two competing reactions: these rate constants may in turn depend on the excitation energy possessed by $A^{\ddagger}$, and will certainly depend on the heats of formation of *all* the products. Calculations involving these and other parameters are the basis of the so-called *quasi-equilibrium* theory (Q.E.T.) of mass spectrometry.

It is misleading to use the intensity of an ion peak as a simple measure of the importance of a particular fragmentation route, unless it is certain that the ion cannot be produced by another route. At low resolution there is the additional complication that the peak may be associated with two ions of equal mass (such as $C_3H_7{}^+$ and $CH_3\overset{+}{C}O$, at m/e 43).

5.7 FRAGMENTATIONS ASSOCIATED WITH FUNCTIONAL GROUPS

In simple monofunctional compounds we can with reasonable certainty predict that the mass spectrum will bear a compound relationship to (a) the nature of the carbon skeleton (whether alkane, alkene, aromatic, etc.) and (b) the nature of the functional group. For difunctional compounds, a more complex interaction can be expected, depending on the relative powers of the two functional groups to direct the fragmentation.

It is always easier to rationalise the mass spectrum of a known structure, than to deduce the structure of an unknown compound from its mass spectrum; other spectroscopic evidence for the presence of functional groups should ideally always be available.

5.7.1 ALKANES AND ALKANE GROUPS

The molecular ion will normally be seen in the mass spectra of the lower *n*-alkanes, but its intensity falls off with increased size and branching of the chain.

Figure 5.6 shows the typical mass spectrum for an *n*-alkane—dodecane in this case. The relative abundances of the ions are also typical, showing maximum abundance around $C_3H_7{}^+$ and $C_4H_9{}^+$, with a weak $M^{\ddagger}$ peak.

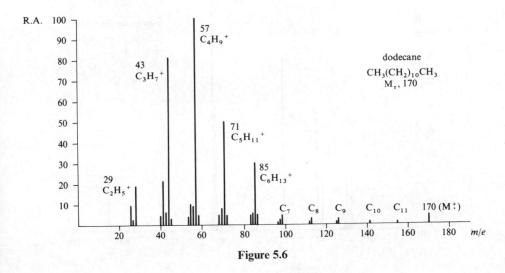

Figure 5.6

Although *n*-dodecane is unbranched, the alkane ions from $C_4H_9{}^+$ up re-arrange in the mass spectrometer to the branched-chain form: this is quite analogous to the Wagner–Meerwein shifts which occur in (for example) $S_N 1$ reactions involving cations, the more stable branched structures being preferred.

Associated with each of these C_nH_{2n+1} ions (from C_2 to C_5) are lesser amounts of the corresponding alkenyl ion (C_nH_{2n-1}) formed by loss of two hydrogen atoms: they appear at m/e 27, 41, 55, 69. Loss of one hydrogen is also seen. Metastable ions of very low intensity can be detected for the fragmentations in which the alkyl ions extrude a smaller molecule: thus $C_2H_5{}^+$ and $C_3H_7{}^+$ extrude H_2: $C_4H_9{}^+$ extrudes CH_4: $C_5H_{11}{}^+$ and $C_6H_{13}{}^+$ extrude C_2H_4: $C_6H_{13}{}^+$ and $C_7H_{15}{}^+$ extrude C_3H_6. These are of too low abundance to be shown in figure 5.6.

Branched-chain alkanes rupture predominantly at the branching points, and the largest group attached to the branching point is often preferentially expelled as a radical. The normal rules of chemistry hold, in that the pre-ference is for the formation of tertiary over secondary over primary cations.

We might predict therefore in the mass spectrum of 2,2-dimethylpentane (a) no molecular ion peak, (b) substantial peaks for $C_3H_7{}^+$, $C_4H_9{}^+$ (because of their inherent stability), (c) a much higher abundance of $C_5H_{11}{}^+$ than would be seen for the fragmentation of n-C_7H_{16} (because of the ease of C_2H_5 expulsion) and (d) a reasonable peak for $C_6H_{13}{}^+$ due to $M - CH_3$.

5.7.2 CYCLOALKANES

Complex fragmentations usually occur for cycloalkanes, ring size obviously being important in relation to ion stability. Typically for the simple members

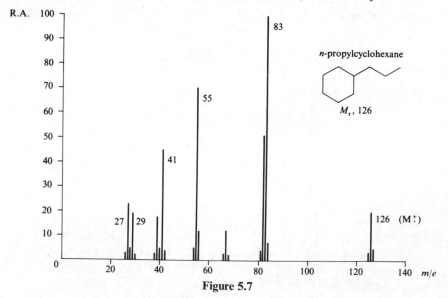

Figure 5.7

the molecular ion peak will be easily seen, its intensity reducing as branching increases.

Common fragmentations are loss of alkenes or alkenyl ions and the splitting off of the side-chains with charge retention by the ring remnant; side-chain loss is simply a special case of fragmentation at a branching point.

Figure 5.7 is the mass spectrum of *n*-propylcyclohexane, which should be interpreted with these points in mind.

5.7.3   ALKENES AND ALKENE GROUPS

Molecular ion peaks for simple alkenes are normally distinctly seen. The commonest fragmentation in alkene groups involves rupture of the allylic bond (β to the double bond), which gives rise to the resonance-stabilised allylic cation. Since the allylic radical is also stabilised, the fragmentation may give rise to peaks corresponding to charge retention by either of the fragments.

$$\text{——CH=CH—CH}_2\text{—R}$$

$$-e \qquad \text{or} \qquad -e$$

$$\text{——CH}^{+\cdot}\text{CH—CH}_2\text{—R} \qquad\qquad \text{——CH}^{+\cdot}\text{CH—CH}_2\text{—R}$$

$$\text{——}\overset{+}{\text{C}}\text{H—CH=CH}_2 + \text{R}\cdot \quad \text{or} \quad \text{——}\overset{\cdot}{\text{C}}\text{H—CH=CH}_2 + \text{R}^{+}$$

$$\text{——CH=CH—}\overset{+}{\text{C}}\text{H}_2 \qquad\qquad \text{——CH=CH—}\overset{\cdot}{\text{C}}\text{H}_2$$

A McLafferty rearrangement (see section 5.6.2) may occur provided the γ-carbon has hydrogen on it.

5.7.4   CYCLOALKENES

Fragmentation of cycloalkenes is directed by the double bond and by the nature of any acyclic alkane residues present, so that allylic rupture and McLafferty rearrangements are common.

In addition, cyclohexene derivatives give the important retro-Diels–Alder reaction discussed in section 5.6.2.

5.7.5 ALKYNES

No simple pattern emerges for the fragmentation of alkynes, which can be applied to complex molecules. Thus for 1-butyne and 2-butyne the molecular ion peak is the base peak, but the molecular ion peak for higher members is weak. Loss of alkyl radicals gives prominent peaks in many cases (at $M - 15$, $M - 29$, etc.) and extrusion of alkenes may give $M - 28$ and $M - 42$ peaks, etc.

5.7.6 AROMATIC HYDROCARBON GROUPS

The molecular ions of aromatic hydrocarbons are always abundant, and M^{+} is commonly the base peak; accordingly, $M + 1$ and $M + 2$ peaks are easily seen. Polynuclear aromatic hydrocarbons have particularly stable molecular ions and doubly or triply charged ions are possible. Doubly charged molecular ions, $m/2e$, will appear at integer m/e values; for example for naphthalene $(C_{10}H_8)$ M^{+} is at m/e 128 and $m/2e$ is at m/e 64. The corresponding $(M + 1)/2e$ peak must, however, appear at noninteger m/e values (m/e 64.5 for naphthalene), which makes it easy to pinpoint the presence of doubly charged ions. Triply charged ions must appear at noninteger m/e values: for naphthalene, $m/3e$ is at m/e 42.6.

Alkylbenzenes, I, are the commonest hydrocarbons in this class; here, the dominant fragmentation is at the benzylic bond, for reasons analogous to those discussed under allylic fission in alkenes.

The stable benzyl cation II would certainly explain the abundant m/e 91 peak seen in the mass spectra of all compounds of this type, but structure II cannot explain the randomisation of C and H atoms shown to occur by isotope substitution studies (see section 55.3). The best explanation is that the ion m/e 91 is the equally stable tropylium ion III; the C_7 ring must be formed by rearrangement of the benzyl group *either before or immediately after* expulsion of the m/e 91 fragment.

The m/e 91 ion subsequently expels C_2H_2 (acetylene) giving m/e 65, which may have the stable structure IV, and this fragmentation gives rise to a metastable ion at $65^2/91 = m/e$ 46.4.

McLafferty rearrangements are observed in alkylbenzenes provided the side-chain has hydrogen on the γ carbon atom. In monoalkylbenzenes this gives rise to an ion at m/e 92, which may be confused with the $m + 1$ peak for the tropylium ion; in this latter case the intensity of m/e 92 would be ≈ 7.7 per cent (7 × 1.1 per cent) of the tropylium ion intensity.

A feasible structure for the m/e 92 ion is the methylenecyclohexadienyl radical ion shown below.

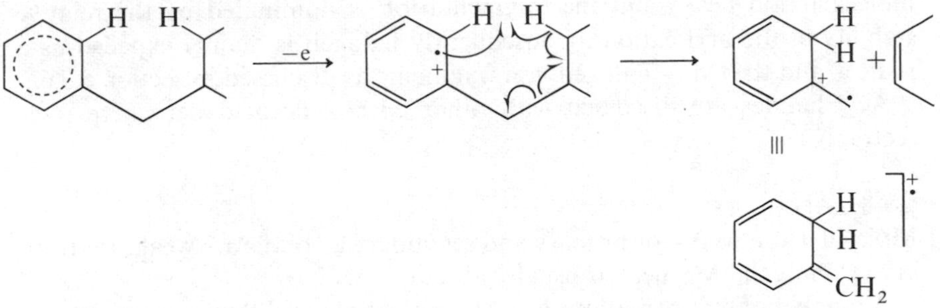

m/e 92
methylenecyclohexadienyl
radical ion

The phenyl cation, $C_6H_5{}^+$, at m/e 77 is produced from many aromatics by rupture of the bond α to the ring, and this ion extrudes C_2H_2 (acetylene) to give m/e 51. This fragmentation gives rise to a metastable ion at $51^2/77 = m/e$ 33.8.

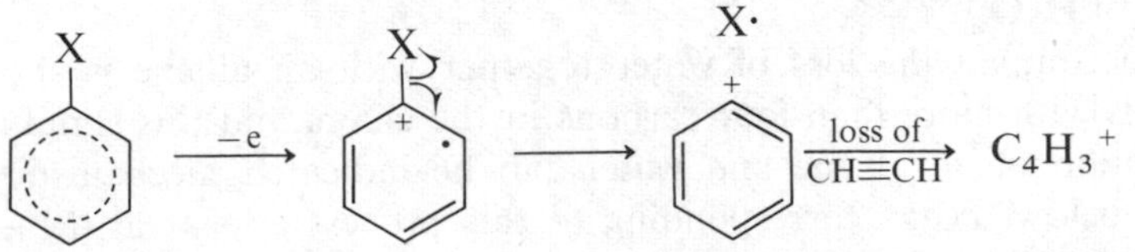

m/e 77 m/e 51

5.7.7 HALIDES

The appearance of the mass spectrum of a halogen-containing compound is profoundly affected by the number of halogen atoms present because of isotopic abundances (as discussed in section 5.3). Fluorine and iodine, being monoisotopic, present no problems in this respect.

The fragmentation of mixed halogen compounds is very complicated, and we shall deal here with compounds containing only one of the halogens.

Aliphatic fluorine compounds, apart from fragmentations appropriate to the alkyl chain, show principally a peak at M − HF (M − 20).

Aliphatic chlorine compounds fragment mainly by loss of HCl (giving two peaks at M − 36 and M − 38): HCl⁺ peaks may also be seen at m/e 36 and 38.

Loss of chlorine as Cl⁺ or Cl· gives rise to low-abundance peaks at m/e 35 and 37, and at M − 35 and M − 37.

Aliphatic bromine compounds fragment similarly to chloro compounds, but loss of Br· is the preferred fragmentation, giving abundant M − 79 and M − 81 peaks.

Aliphatic iodine compounds show peaks corresponding to I⁺ (m/e 127), M − I (M − 127), and M − H₂I (M − 129).

Aryl halides, with halogen directly attached to the ring, show abundant molecular ion peaks, but the fragmentation is dominated by the relative stability of the aryl cations. Consequently halogen is mainly expelled as a radical, and the (M − halogen) ion fragments as discussed in section 5.7.6.

Acyl halides are discussed with other carboxylic acid derivatives: see section 5.7.18.

5.7.8 ALCOHOLS

Molecular-ion peaks for primary and secondary alcohols are weak; for tertiary alcohols the M⁺ peak is usually absent.

A number of fragmentations is open to alcohols, and their relative importance depends on whether the alcohol is primary, secondary, tertiary, aliphatic or aromatic; the most important fragmentation is normally rupture of the bond β to oxygen.

Alcohols of low volatility can be converted into their trimethylsilyl ethers (ROH → ROSiMe₃), which are much more volatile because of the loss of hydrogen bonding present in the alcohols themselves (see section 5S.4).

Primary aliphatic alcohols (*alkanols*) show M − 18 peaks corresponding to loss of H₂O.

An accompanying loss of water together with an alkene is shown by alcohols with more than four carbons in the chain, and this simultaneous elimination of an alkene and water can be indicated mechanistically as shown below. Peaks corresponding to this process appear at (M − 18 − C_nH_{2n}), for example (M − 18 − 28) for loss of water and C₂H₂, etc.

Long-chain members may show peaks corresponding to successive loss of H radicals at M − 1, M − 2 and M − 3; this can be represented as shown below.

$$\begin{array}{cccc} \overset{H}{\underset{\underset{H}{|}}{-\overset{|}{C}}}-\overset{+}{\ddot{O}}-H & \longrightarrow & -\overset{H}{\underset{}{C}}=\overset{+}{\ddot{O}}H & \longrightarrow & -\overset{H}{\underset{}{C}}=\overset{+}{\ddot{O}} & \longrightarrow & -C\equiv\overset{+}{O}\cdot \\ M^{+} & & M-1 & & M-2 & & M-3 \end{array}$$

Secondary and tertiary aliphatic alcohols preferentially fragment by loss of alkyl radicals, the ease of elimination increasing with increased size and branching in the radical. Thus the alcohol shown would be expected to give rise to a prominent peak at M − C_4H_9, with less abundant peaks at M − C_3H_7 and M − C_2H_5.

$$C_4H_9-\underset{\underset{C_2H_5}{|}}{\overset{\overset{+}{O}H}{\overset{|}{C}}}-C_3H_7 \longrightarrow C_4H_9\cdot + \underset{\underset{C_2H_5}{|}}{\overset{\overset{+}{O}H}{\overset{\|}{C}}}-C_3H_7$$

Aromatic alcohols with the OH group in the benzylic position fragment so as to favour charge retention by the aryl group; thus in 1-phenylethanol the base peak corresponds to elimination of $CH_3\cdot$. Peaks corresponding to $ArCO^+$ and Ar^+ are also shown. The peak at m/e 107 given by these alcohols (the base peak in the case of 1-phenylethanol) is best represented as the hydroxytropylium ion. Loss of CO from this ion gives m/e 79, and loss of H leads on to the phenyl cation at m/e 77, which loses C_2H_2 to give m/e 51.

$$\begin{array}{c} OH \\ | \\ \text{⬡}-CHCH_3 \end{array}$$

1-phenylethanol

$$\begin{array}{ccccccc} \overset{OH}{\underset{}{\text{⬡}^+}} & \xrightarrow{-CO} & \overset{H\;\;H}{\underset{}{\text{⬡}^+}} & \longrightarrow & C_6H_5{}^+ & \longrightarrow & C_4H_3{}^+ \\ m/e\ 107 & & m/e\ 79 & & m/e\ 77 & & m/e\ 51 \\ \text{hydroxytropylium} \\ \text{ion} \end{array}$$

5.7.9　PHENOLS ·

Simple phenols give strong molecular ion peaks.

　　The commonest fragmentation is loss of CO (M − 28) and CHO (M − 29), which can only be represented as shown below.

Phenols with alkyl side-chains also undergo benzylic fission, leaving variants of the hydroxytropylium ion (see section 5.7.8).

$$m/e\ 107\ when\ X = H$$

5.7.10　ETHERS, ACETALS AND KETALS

On the whole, molecular ion peaks for these classes are weak.

　　Aliphatic ethers principally fragment at the bond β to oxygen, and the largest group is expelled preferentially as a radical (compare alcohols). Peaks appear therefore at M − CH_3, M − C_2H_5, etc.

Where β hydrogen is present, the oxonium ion may fragment further to eliminate an alkene.

　　Fission of the C—O bond (the α bond) occurs to a small extent, charge retention by carbon being favoured over retention by oxygen (see page 213).

　　Acetals and ketals show simple extensions of the ether fragmentation processes, rupture of the β bond being favoured over the α.

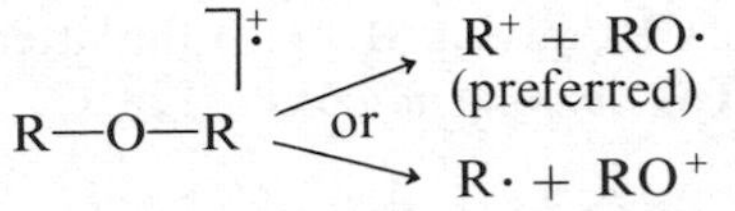

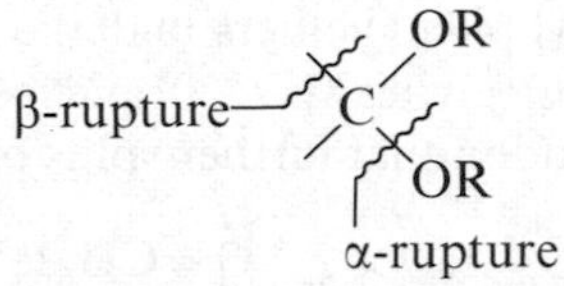

R—O—R $\begin{array}{c}\\\text{or}\end{array}$ $\left.\begin{array}{l} R^+ + RO\cdot \\ \text{(preferred)} \\[1em] R\cdot + RO^+ \end{array}\right.$ β-rupture

α-rupture

In addition, cyclic acetals such as ethylene ketals will fragment to resonance-stabilised cyclic oxonium ions: so easily are these formed that they often dominate the fragmentation pattern and give rise to the base peak.

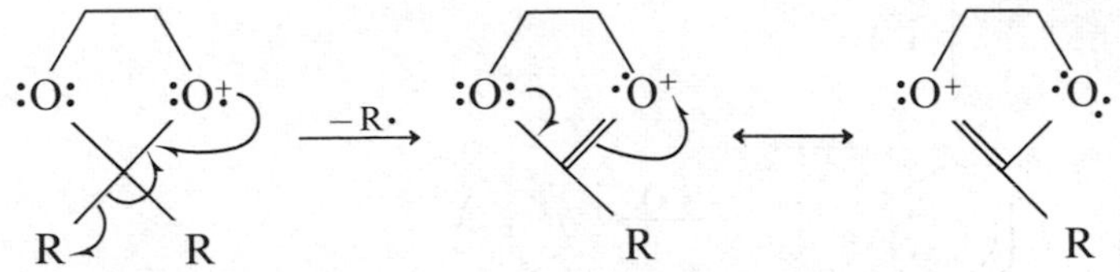

m/e 87, 101, etc.

For cyclic ketals in which the ketone residue is itself cyclic (as in the ethylene ketals of cyclohexanone derivatives) the initial bond rupture is modified by a second rupture so that an alkonium alkene ion is formed.

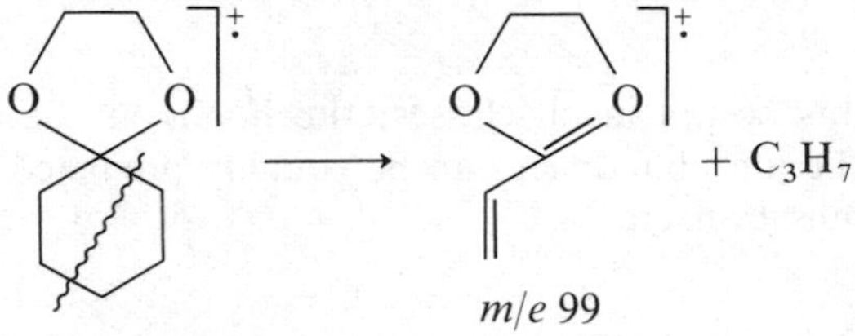

$+ C_3H_7\cdot$

m/e 99

Forming the ethylene ketal constitutes a valuable derivatisation technique for directing the fragmentation of ketones (see section 5S.4).

Aromatic ethers other than methyl ethers commonly fragment by variants of the β-hydrogen transfer discussed above for aliphatic ethers: a peak at m/e 94 is formed, which extrudes CO giving m/e 66.

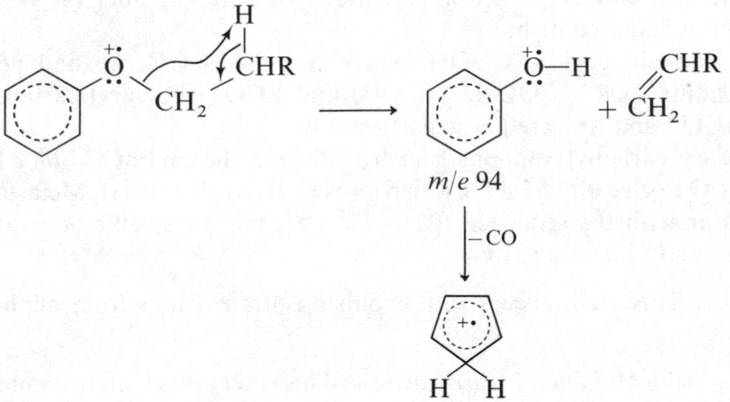

m/e 94

$-$CO

m/e 66

Methyl phenyl ethers undergo two main fissions at the C—O bonds: loss of HCHO (giving M − 30) and loss of CH_3 (giving M − 15), the latter process giving an ion that further splits out CO, leaving m/e 65.

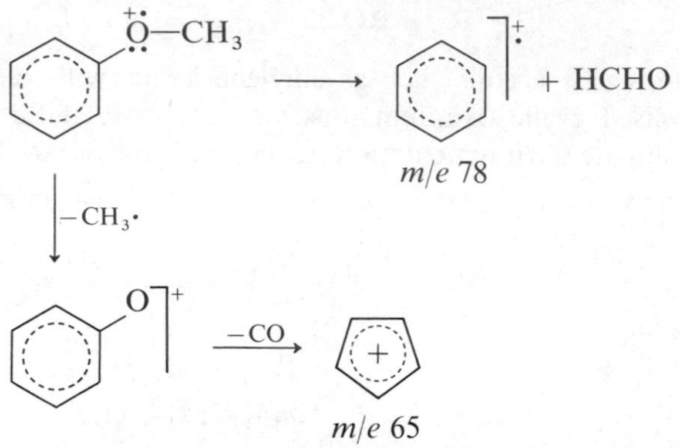

m/e 78

m/e 65

5.7.11 CARBONYL COMPOUNDS GENERALLY

Before discussing the idiosyncratic behaviour of individual carbonyl classes, we should draw attention to the important similarities in their fragmentation modes.

α-Cleavage. This occurs in all classes; the bonds at the carbonyl group rupture, and the ion abundance can be roughly predicted on the basis of resonance stabilisation, etc.

In summary

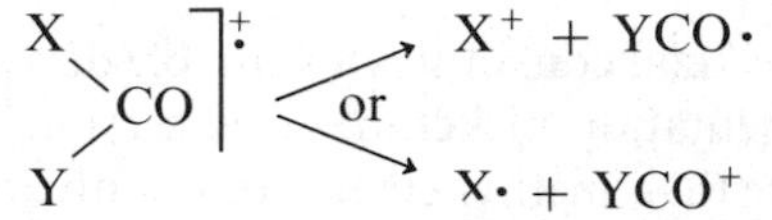

Thus in RCHO aldehydes, where X or Y is H, we find peaks corresponding to R^+, RCO^+ and HCO^+. For aryl aldehydes, the stability of $ArCO^+$ makes this a major contributor.

For carboxylic acids (RCOOH) where X or Y is OH, we find peaks corresponding to R^+, $COOH^+$ (m/e 45) and RCO^+. For aryl carboxylic acids, $ArCO^+$ and Ar^+ are particularly stable.

For all aryl carbonyl compounds such as PhCOX the ion $PhCO^+$ (m/e 105) will lose CO to give Ph^+ (m/e 77), which loses C_2H_2 to give m/e 51. Metastable ions appear with the sequence $105 \rightarrow 77 \rightarrow 51$, the metastable m/e values being 56.5 and 33.8, respectively.

β-Cleavage. This often occurs with expulsion of alkyl ions from aliphatic aldehydes.

β-Cleavage with McLafferty rearrangement. This is very much more common, provided γ-hydrogen is present (see over and section 5.6.2).

For the simple unbranched aliphatic aldehydes, X = H and R = H, and we can expect an ion at m/e 44 for this fragmentation. For the corresponding ketones, X must be CH_3, C_2H_5, etc., giving peaks at m/e 58, 72, etc. For carboxylic acids, with X = OH, the peaks appear at m/e 60, 74, etc.

We can now go on to the individual carbonyl classes, noting common ion values and any fragmentation in addition to α and β-cleavage.

5.7.12   ALDEHYDES

Aliphatic aldehydes give weak molecular ion peaks, whereas aryl aldehydes give strong $M^{\ddagger}$ peaks.

Loss of H$\cdot$ from the molecular ion is particularly favoured by aryl aldehydes because of the stability of $ArCO^+$; nevertheless M − 1 peaks are prominent in the mass spectra of all aldehydes, whether aliphatic or aromatic. α-Cleavage gives R^+ peaks and the HCO^+ ion at m/e 29: $C_2H_5^+$ also appears at m/e 29, and for higher aliphatic aldehydes $C_2H_5^+$ is more likely to be the source of m/e 29 than HCO^+.

Other cleavages are discussed under carbonyl compounds in section 5.7.11.

5.7.13   KETONES AND QUINONES

Molecular ion peaks for all ketones are usually strong. Most of the abundant ions in the mass spectra of ketones can be accounted for by α-cleavages and McLafferty rearrangements, and indeed the base peak for methyl ketones and phenyl ketones is frequently CH_3CO^+ (m/e 43) and $PhCO^+$ (m/e 105), respectively: $PhCO^+$ fragments as usual to m/e 77 and m/e 51.

When RCOR' undergoes α-cleavage the larger group is preferentially expelled as the radical, with concomitant charge retention by the smaller version of RCO^+.

Common values for the McLafferty rearrangement ions of aliphatic ketones are m/e 58, 72, 86, etc., but the higher members of the series may undergo a second elimination of an alkene to generate a further series of ions.

For example m/e 86 may lose ethylene giving an ion at m/e 86 − 28, and loss of C_3H_6 will correspond to loss of 42 mass units. For simple aromatic ketones the McLafferty rearrangement ions might arise at m/e 120, 134, etc. All of the McLafferty-rearrangement ions appear at even m/e values.

Ethylene ketal derivatives of ketones (mentioned on page 213) have very strongly directed fragmentations, and are discussed further in section 5S.4.

Quinones mainly undergo α-cleavage, both α bonds being capable of rupture.

p-benzoquinone   α-naphthoquinone   9,10-anthraquinone

In *p*-benzoquinone and naphthoquinone these fragmentations lead to peaks at m/e 54 and at M − 54, whereas α-cleavage in anthraquinone gives the whole range of ions corresponding to M − CO, M − 2CO and m/e 76 $(C_6H_4^+)$.

5.7.14   CARBOXYLIC ACIDS
Molecular ion peaks are usually observable, but weak. Common peaks arising from α-cleavage and McLafferty rearrangements have already been mentioned in section 5.7.11.

5.7.15   ESTERS
Methyl esters are usually more convenient to study than the free acids, because they are more volatile. The molecular ion peak is weak but discernible in most cases, and the fragmentation is a mixture of α-cleavage and (where appropriate) McLafferty rearrangements. Common ion values for α-cleavage are therefore RCO^+, R^+, CH_3O^+ (m/e 31) and CH_3OCO^+ (m/e 59); the McLafferty ion at m/e 74 $(CH_2{=}C(OH)OCH_3)^{\ddagger}$ is the base peak in the saturated straight-chain methyl esters from C_6 to C_{26}.

Higher esters are complicated by the possibility of two fragmentation modes—fragmentations of the acyl group (RCO—) and of the alkyloxy group (ROCO—). The acyl group fragmentation is a simple extension of that discussed under methyl esters, but where the alkyloxy group is C_2H_5O (ethyl esters) or higher, the McLafferty ion (m/e 88 for ethyl esters) undergoes both loss of an alkene and an alkenyl radical to give ions at m/e 60 and 61 $(CH_3CO_2H^+$ and $CH_3CO_2H_2^+)$. The number of possible esters is extremely high, and the relative importance of the different fragmentation modes is almost unique for each member, so that we must be content with these general pointers.

5.7.16 AMIDES

Primary aliphatic amides, $RCONH_2$, undergo α-cleavage to R· and $CONH_2^+$ (m/e 44). Where possible, McLafferty rearrangement occurs to homologous variants of CH_2=C(OH)$NH_2^{\ddagger}$ at m/e 59, 73, etc. Loss of NH_2 gives M − 16 peaks.

Primary aryl amides usually have this as their primary fragmentation, leading to $ArCO^+$ ions.

For secondary (RCONHR′) and tertiary amides (RCONR′R″) the number of individual variations is enormous as in esters, when we consider that fragmentation must take into account the nature of the three groups R, R′ and R″.

5.7.17 ANHYDRIDES

Molecular ion peaks are weak or absent.

Saturated acyclic anhydrides (acetic anhydride, etc.) fragment mainly to RCO^+ (with m/e values 43, 57, etc.), although chain-branching may cause such easy fragmentation that the complete RCO^+ is not detected. Ions at M − 60 and m/e 60 are common (CH_3CO_2H = 60), as are m/e 42 (CH_2=CO^+) and McLafferty ions at m/e 44, 58, etc.

Cyclic aliphatic anhydrides such as succinic anhydride show a strong or base peak at M − 72 caused by loss of CO_2 + CO from $M^{\ddagger}$, and this is also shown by the cyclic aromatic anhydrides (phthalic anhydride, etc.). For these latter, other common ions are $ArCO^+$, $ArCO_2H^+$ and M − CO: loss of H from $ArCO^+$ can also occur.

5.7.18 ACID CHLORIDES

Aliphatic members show fragmentations associated with the Cl and CO groups, so that the following ions are common: HCl^+, M − Cl, $COCl^+$, RCO^+, etc. The effect of isotope abundance (^{35}Cl: ^{37}Cl ≈ 3:1) makes it easy to identify the chlorine-containing peaks. For aryl acid chlorides (for example benzoyl chloride) the stable $ArCO^+$ makes loss of Cl· from $M^{\ddagger}$ a dominant process.

5.7.19 NITRILES

Molecular ion peaks are usually weak or absent, although an M − 1 ion (R—CH=C=N^+) may be seen.

For the lower aliphatic members M − 27 (corresponding to M − HCN) is seen, but from C_4 on the McLafferty ion is frequently the base peak: these ions appear at m/e 41, or 55, or 69, etc., and can be represented as the homologues of CH_2=C=$NH^{\ddagger}$.

Aryl nitriles often show M − HCN peaks, although if alkyl side-chains are present benzylic rupture will give rise to the main series of ions.

5.7.20 Nitro Compounds

Molecular ion peaks for aliphatic members are usually absent, while for many aromatic nitro compounds the M^+ peak is strong.

The mass spectra of aliphatic nitro compounds mainly correspond to fragmentation of the alkyl chain, although peaks for NO^+ (m/e 30) and NO_2^+ (m/e 46) appear also.

For aryl nitro compounds, peaks corresponding to NO^+ (m/e 30), NO_2^+ (m/e 46), $M - NO$ ($M - 30$) and $M - NO_2$ ($M - 46$) all appear commonly, and successive loss of NO and CO ($M - 58$) is also observed. Loss of NO from the molecular ion leaves an ion of structure ArO^+, which can only arise if $ArNO_2^{+}$ rearranges in the spectrometer.

Ortho effects are frequently encountered in aryl nitro compounds where a group *ortho* to NO_2 contains hydrogen: peaks corresponding to $M - OH$ ($M - 17$) appear, indicating loss of oxygen from NO_2 together with H from the *ortho* substituent.

5.7.21 Amines and Nitrogen Heterocycles

An odd number of nitrogen atoms in the molecule means an odd relative molecular mass (molecular weight).

For primary aliphatic amines, the base peak is $CH_2{=}\overset{+}{N}H_2$ (at m/e 30) formed by expulsion of a radical from M^{+}. Higher homologues may also appear at m/e 44, 58, etc., but these are less abundant ions. Loss of an alkene (for example, loss of C_2H_4) may give peaks at $M - 28$, etc.

Secondary and tertiary amines behave analogously, and loss of the substituent alkyl radicals is also observed.

Primary aryl amines principally fragment by loss of H ($M - 1$) and HCN ($M - 27$): thus aniline gives rise to an ion at m/e 66 ($93 - 27$) whose structure can be represented as the cyclopentadienyl radical cation. For N-alkyl-anilines, α-cleavage of the alkyl group is common.

Aromatic heterocyclic bases give rise to abundant M^+ peaks, and for pyridine and quinoline the principal fragmentation is loss of HCN ($M - 27$). Alkyl-substituted derivatives fragment at the 'benzylic' bond as for alkyl-benzenes, and loss of HCN commonly follows on this process.

5.7.22 Sulphur Compounds

Fragmentations of thiols (mercaptans and thiophenols) and sulphides bear close comparison to their oxygen analogues, with the additional complication due to $M + 2$ and $m + 2$ peaks because of ^{32}S and ^{34}S isotope abundances. Normally, molecular ion peaks are clearly seen.

Aliphatic thiols may also give rise to the following ions: S^+, HS^+, H_2S^+ and $M - H_2S$.

The commonest sulphur heterocycles are thiophen derivatives: thiophen itself gives a strong M^+ peak, together with HCS^+ (the thioformyl ion) and $M - HCS$. As for Ph^+, loss of $CH{\equiv}CH$ produces a peak at $M - 26$.

For sulphides (RSR') α-cleavage gives RS^+, often followed by loss of CS to give $RS - CS$ peaks. Ions also arise due to $M - HS$ processes.

SUPPLEMENT 5

5S.1 ALTERNATIVES TO ELECTRON-IMPACT IONISATION

The ability to measure accurately the relative molecular mass (molecular weight) of organic compounds by mass spectroscopy is only possible if a sufficiently stable molecular ion can be formed, and we have seen that many classes of compound do not do so when electron-impact ionisation is used. Partly, the reason lies in the large amount of excess energy imparted to the molecular ion by 70 eV bombardment; not only does this lead to rapid decomposition of many molecular ions, but very complex fragmentation patterns often result. Two useful alternatives to electron-impact ionisation are worth noting, each of which goes some way toward complementing the data obtained from conventional 70 eV spectra.

5S.1.1 Chemical ionisation

This is brought about by mixing the sample (at $1.3 \times 10^{-2} \, N \, m^{-2} \equiv 10^{-4}$ torr) with a reactant gas (at $1.3 \times 10^2 \, N \, m^{-2} \equiv 1$ torr) and submitting this mixture to electron bombardment. The reactant gas most commonly used is methane, although other hydrocarbons have also been used: on electron impact, it is the methane which is ionised, and two ensuing ion–molecule reactions are important.

$$CH_4^{+\cdot} + CH_4 \longrightarrow CH_5^+ + CH_3 \cdot$$
$$CH_3^+ + CH_4 \longrightarrow C_2H_5^+ + H_2$$

The CH_5^+ and $C_2H_5^+$ ions then react with sample molecules, inducing them to ionise, and these ions are separated magnetically and electrostatically in the normal way. Unfortunately, CH_5^+ and $C_2H_5^+$ do not react with all classes of organic compound in the same way: for n-alkanes the base peak is normally the $M - 1$ peak at $C_nH_{2n+1}^+$, whereas for many basic compounds (amines, alkaloids, amino acids) the base peak is the $M + 1$ peak. The $M + 1$ peaks arise by protonation of nitrogen, and for the alkanes the $M - 1$ peaks can be explained by the following reactions.

$$C_nH_{2n+2} + CH_5^+ \longrightarrow C_nH_{2n+1}^+ + CH_4 + H_2$$
$$C_nH_{2n+2} + C_2H_5^+ \longrightarrow C_nH_{2n+1}^+ + C_2H_6$$

The principal advantages of chemical ionisation over electron impact are: (a) more abundant peaks related to the molecular ion, whether M^+ or $M + 1$ or $M - 1$; (b) simpler fragmentation patterns, which make it easier in many cases to study the kinetics of reaction of individual ions; (c) easy application of gas chromatography–mass spectroscopy interfacing, since methane can be used not only as reactant gas (in the chemical ionisation) but also as the carrier gas in the gas chromatograph (see section 5S.2).

5S.1.2   Field ionisation.

An organic compound in the gas phase can be ionised when the molecules pass near a sharp metal anode carrying an electric field of the order 10^{10} V m^{-1}. Electrons are 'sucked' from the sample molecules into incomplete orbitals in the metal, and the resulting molecular ions are then repelled towards a slit cathode. Primary focusing takes place at the cathode slit before the ions pass through the entrance slit of the mass spectrometer to be focused magnetically and electrostatically as in electron-impact studies.

As in the case of chemical ionisation, the principal advantages of field ionisation from an organic chemist's point of view are the increased abundance of molecular ions and the minimisation of complex fragmentations and rearrangements. Disadvantages are the lower sensitivity and resolution obtained.

Outstanding advantages can be achieved by a modification of the technique in which the sample is deposited directly onto the anode, and the high field produces not only ionisation but desorption. Unstable and involatile material can be handled in this way, and molecular ion peaks have thus been produced from complex naturally occurring compounds (notably the carbohydrates) that do not show M$^+$ peaks on electron impact.

5S.2   GAS CHROMATOGRAPHY – MASS SPECTROSCOPY (g.c./m.s.)

The supreme analytical power of gas chromatography can be coupled with the structure probing of the mass spectrometer and the combined technique (gas chromatography–mass spectroscopy or g.c./m.s.) used diversely to investigate complex reaction products, or naturally occurring mixtures, or the organic content of fossil rocks. Provided the material is capable of being gas chromatographed (for which the prime requisites are moderate vapour pressure and reasonable stability in the gas phase), and provided that a sufficient quantity of each chromatographed component is available, each peak from the gas chromatograph can be bled to the mass spectrometer in succession.

Two major instrumental problems have to be solved before g.c./m.s. becomes semiroutine. Firstly, the rate at which peaks are eluted from the g.c. column demands very fast scan speeds from the mass spectrometer. Modern instruments meet this requirement and obviate the earlier need either to condense chromatographed components out of the carrier gas stream (not easily successful with minor constituents), or to stop the chromatograph while each peak is having its mass spectrum recorded. Secondly, the components being eluted from the chromatograph are present at high dilution in the carrier gas, and the total eluant cannot be fed directly into the low-pressure system of the mass spectrometer (unless chemical ionisation is being used, as discussed in section 5.S.1.1). Various devices have been developed to overcome this problem, all of them dependent for their success on the higher diffusion rate of the carrier gas (H_2 or He) compared to the sample molecules: for example in the Watson–Biemann separator the total eluant is passed through a porous glass tube, which is surrounded by an evacuated compartment, and carrier gas is pumped off preferentially. The eluant, enriched typically several hundredfold, is then bled through a sinter into the spectrometer.

When a fast-scanning mass spectrometer is dealing with a succession of peaks rapidly eluting from a gas chromatograph, the data output of the mass spectrometer is enormous; the only satisfactory treatment of such data is to interface the mass spectrometer to a computer storage facility, and to collect the entire output for later retrieval and digestion. With such data-acquisition and handling it is possible to analyse and correlate the mass spectra of individual constituents in the chromatogram. The computer can be programmed to print out directly the element compositions of parent and fragment ions, and to search out those constituents that share a common fragment ion, etc. For example, it is possible to ask the computer to print out a list of those constituents that have m/e 91 or have $M - 18$ peaks in their mass spectra.

The application of g.c./m.s. impinges on any area where complex mixtures have to be separated and identified in small quantities: biological material is a rich source of such work, and research in carbohydrates, proteins, fats and steroids have all benefited from its power.

5S.3   Isotope Substitution in Mass Spectroscopy
The incorporation of less-abundant isotopes into organic molecules makes it relatively easy to follow by mass spectroscopy the mechanisms of a number of reactions, both in conventional laboratory or biological sequences and in the fragmentations in the mass spectrometer itself. A few examples will be given to illustrate the scope of the technique, but the variations are too numerous to list.

The isotopes that might be used include ^{2}H, ^{13}C, ^{18}O, ^{15}N, ^{37}Cl, etc., but the cost of enriched-isotope sources is so high that most work has been done on the cheapest (^{2}H) and comparatively less on ^{13}C and ^{18}O. To be certain of a mechanistic step in a reaction we must often be able to show nearly 100 per cent inclusion of the isotope label in the product of the reaction (or 100 per cent exclusion), since values substantially less than 100 per cent could be associated with scrambling of atoms *within* the mass spectrometer itself.

An early successful application of isotope labelling was the investigation of ester hydrolysis, which could conceivably involve acyl–oxygen fission or alkyl–oxygen fission (or a mixture of both). Labelled with ^{18}O as shown, acyl–oxygen fission leads to ^{18}O incorporation in the alcohol, while alkyl–oxygen fission leads to incorporation in the acid. (In general the former is observed, except for esters of tertiary alcohols.) The distinction is easily made by mass spectroscopy, since the acid and alcohol can be isolated and

$$R-\overset{\overset{\displaystyle ^{16}O}{\|}}{C}-^{18}O-R' \left\langle \begin{array}{l} \overset{H_2O}{\longrightarrow} \\ \text{or} \\ \overset{H_2O}{\searrow} \end{array} \right. \begin{array}{l} R-\overset{\overset{\displaystyle ^{16}O}{\|}}{C}-^{16}OH + H-^{18}OR' \quad \text{acyl–oxygen fission} \\[2em] R-\overset{\overset{\displaystyle ^{16}O}{\|}}{C}-^{18}OH + H-^{16}OR' \quad \text{alkyl–oxygen fission} \end{array}$$

the $M^{\ddagger}$ peak for each measured: ^{18}O incorporation gives the $M^{\ddagger}$ peak two mass units higher than the 'normal' ^{16}O analogue.

In structural organic chemistry it is often desirable to know how many enolisable hydrogens are adjacent to a carbonyl group. The replacement of —CH_2CO— by —CD_2CO— can be executed rapidly by treating —CH_2CO— with D_2O and base, and subsequent measurement of $M^{\ddagger}$ will show the degree of deuterium incorporation. This method has been elegantly extended so that in-column deuterium exchange takes place *during* gas chromatography of the material, which can then be studied by the combined g.c./m.s. method.

The biochemical applications of isotope labelling are legion, and to mention any one would perhaps invidiously imply its relative importance: of outstanding interest to organic chemists have been the acetate-labelling studies of biosynthetic pathways. The modes of incorporation of ^{13}C into steroids, carbohydrates, fatty acids, etc. have been fundamental revelations.

In mass spectroscopy itself, solutions to the problems of ion structure and fragmentation mechanisms have been keenly sought. The fragmentations of the toluene molecular ion are archetypal, and illustrate well the complexities involved.

We saw earlier (section 5.7.6) that the toluene molecular ion (m/e 92) gives rise to a series of daughter ions at m/e 91, 65 and 39, and that the ion m/e 91 is best represented as the tropylium ion: the crudest representation of these processes would be as shown below. The evidence which belies this simple interpretation is summarised as follows:

(i) ^2H labelling shows that loss of H· from the toluene molecular ion is almost random, and does not take place exclusively from the benzylic carbon.
(ii) ^{13}C labelling of the side-chain shows that this carbon is not exclusively expelled in the transition m/e 91 → m/e 65.

(iii) double ^{13}C labelling (the methyl carbon and its neighbour) show that although the two ^{13}C atoms are adjacent in the original toluene molecule, complete randomisation occurs somewhere between there and the decomposition of ion m/e 91.

What explanation can be offered? All that is certain is that no simplistic 'mechanism' is valid for this sequence of fragmentations; the postulation of the tropylium ion structure for m/e 91 is certainly better than the benzylic carbonium ion, but even this may not be a true postulate.

5S.4   DERIVATISATION OF FUNCTIONAL GROUPS

Two main benefits may accrue from converting the functional group in a molecule to one of its functional derivatives. The derivative may be more volatile than the parent, or the derivative may give a simpler mass spectrum, possibly also with an enhanced molecular ion peak. A few examples will illustrate the value of the method.

Carbohydrates are very difficult to handle in 70 eV mass spectroscopy since they are very involatile and give no molecular ion peaks. They can be converted rapidly and cleanly into their trimethylsilyl ethers by treatment with a mixture of hexamethyldisilazane ($Me_3SiNHSiMe_3$) and trimethylsilyl chloride (Me_3SiCl): the trimethylsilyl ethers are sufficiently volatile to be easily capable of g.c./m.s. analysis.

The following conversions are widely used to effect similar improvement in other functional classes.

$$RCO_2H \longrightarrow RCO_2Me$$
$$ROH \longrightarrow ROMe \text{ or } ROSiMe_3 \text{ or } ROCOMe$$
$$ArOH \longrightarrow ArOMe \text{ or } ArOSiMe_3 \text{ or } ArOCOMe$$
$$RCONH- \longrightarrow RCONMe-$$

Functional groups do not possess an equal ability to direct molecular fragmentations, since the activation energies for the formation of the fragment ions will be different (depending on the stability of the fragment ions). An approximate ranking order (most strongly directing function last) would be carboxyl, chloride, methyl ester, alcohol, ketone, methyl ether, acetamido, ethylene ketal, amine. We can often make use of this ranking order to simplify the mass spectrum of a molecule, by converting the functional group to a derivative of greater directing propensity. A good example is found in the case of ketones: if a ketone molecular ion gives rise to a large number of fragment ions, we can convert the ketone to the ethylene ketal and usually the simple fragmentation of this group will dominate the spectrum. Ethylene ketals fragment to stable oxonium ions, and the low energy of activation for their formation means that alternative reactions cannot compete. The resulting simplification in the fragmentation pattern makes structural deductions much clearer: in the example shown, the position of the two substituents (Me, Et) in the substituted cyclohexanone can be deduced

$$\text{CH}_2\text{OH} \\ | \\ \text{CH}_2\text{OH}$$

ketone　　　ethylene　　　　　stabilised oxonium ion
　　　　　　　ketal　　　　　　　　　　m/e 99

m/e 99　　　and　　　m/e 113

from the two ions' m/e values, given the known fragmentation modes of the parent cyclohexanone ethylene ketal.

The phenomenon of influencing fragmentation modes by derivatisation is usually termed *directed fragmentation*.

5S.5 ALTERNATIVES TO MAGNETIC/ELECTROSTATIC FOCUSING

We have seen that routine organic mass spectroscopy is taken to mean 70 eV electron-impact ionisation, followed by mass separation on the basis of magnetic (and additionally electrostatic) focusing of positive ions. Three principal alternative means of mass separation have been developed, each of which has advantages and disadvantages, although on balance they have not overtaken magnetic/electrostatic focusing as the method of choice for the greatest number of instruments in current use.

Time-of-flight mass spectroscopy differentiates the positive-ion masses by measuring the times that they take to traverse a flight tube of approximately 1 m length. The ions are first of all generated in short pulses and accelerated to uniform kinetic energy; heavier ions travel more slowly to the collector than light ions, although the very short time differences between ions can be of the order 10^{-8} s. The ion pulses enter the flight tube at intervals of around 10^{-4} s (a frequency of 10 kHz), and the mass scan must be of the same time scale: usually, the spectrometer conditions are optimised by presenting the spectrum on an oscilloscope (also scanning at 10 kHz), and a permanent record is obtained from an analogue recording of the oscilloscope trace. The extremely fast scan time and wide-slit dimensions of time-of-flight spectrometers makes them particularly useful for dealing with transient species, for example those produced in shock tubes or flames: by using rapid photographic techniques (drum cameras, etc.) the oscilloscope trace can be recorded over a very small time span.

Quadrupole mass spectrometers have as their core an ion tube (the 'mass filter') containing four accurately aligned metal rods, arranged symmetrically around the long axis of the tube. Opposite pairs of rods are coupled together, and a complex electric field is set up within the four rods by applying direct current voltage (a few hundred volts) across the coupled rods, superimposed by radiofrequency potential. In operation, positive ions are propelled into the mass filter; the ratio of d.c. voltage to r.f. voltage is chosen, and these voltages scanned (at constant ratio) from zero to a maximum value when they return to zero to repeat the scan. As the d.c. and r.f. voltages are built up, a hyperbolic potential field is established within the four rods: at any particular value for this field (that is, at any particular point in the scan) most ions will be deflected towards the rods and discharged there. For ions of the appropriate mass-to-charge ratio, however, there will be a particular value of the d.c./r.f. scan that will induce the ion to describe a modulated wave-like path along the hyperbolic field. These ions will reach the collector (usually an electron multiplier) and be recorded there.

The entire mass spectrum is obtained by the d.c./r.f. scan through the correct values for each m/e ratio.

Quadrupole mass spectrometers are small and relatively cheap, and are widely used for fast-scan g.c./m.s. work associated with kinetic or pyrolysis experiments in the gas phase. The major limitation in the cheaper versions is low resolution, which restricts their use to the study of relative molecular masses (molecular weights) up to around 500. Within this limit, however, they find wide use in monitoring processes, for example in the analyses of respiratory gases, of atmospheric gases in tanks and spacecraft, and of hazardous components in working atmospheres.

Ion cyclotron resonance spectroscopy involves generation of positive ions by electron impact, after which they are drawn through a short analyser tube by a small static electric field. While in the analyser, a magnetic field (of the order 0.8 T) is superimposed: this causes the ions to perform a series of cycloidal loop-the-loops along the analyser, the angular frequency of which lies in the radiofrequency range (around 300 kHz). This frequency (the *cyclotron resonance frequency*, ω_c) is a function of both the magnetic field strength and the m/e value for the ion ($\omega_c = Be/m$), and the 'mass spectrum' is obtained by scanning the field B until ω_c comes to resonance with a fixed radiofrequency source beamed onto the analyser tube. As in n.m.r., the resonance condition is reached when ω_c equals the frequency of the radiofrequency source, and, at resonance, measurable radiofrequency energy passes from the r.f. circuit to the ion beam.

Ion cyclotron-resonance spectroscopy is relatively new, and its principal application has been in the study of ion–molecule reactions. Although pressures used in the spectrometer are low ($\approx 1.3 \times 10^{-4}\,\mathrm{N\,m^{-2}} \equiv 10^{-6}$ torr), the cycloidal path is long and drift times are long (of the order of milliseconds) so that the probability of ions colliding with molecules is enhanced.

A general example of how the technique can be applied to the investigation of ion structure might be a situation where two molecules A and B both give rise to a daughter ion at m/e x: has the $m/e = x$ ion from A the same structure as the $m/e = x$ ion from B? If we can demonstrate by isotope substitution that the daughter ion from A reacts in the ion cyclotron resonance spectrometer with an admixed neutral molecule to produce a new ion at m/e y, then we can repeat the experiment with the daughter ion of B mixed with the same neutral molecule. If the new ion at m/e y is not produced in this test, then the daughter ions of A and B, even though they have the same m/e value, cannot have the same structure.

FURTHER READING

MAIN TEXTS

F. W. McLafferty, *Interpretation of Mass Spectra*, Benjamin, New York (1966).

H. C. Hill, *Introduction to Mass Spectrometry*, Heyden, London (1966).

K. Biemann, *Mass Spectrometry. Organic Chemical Applications*, McGraw-Hill, New York (1962).

R. A. W. Johnstone, *Mass Spectrometry for Organic Chemists*, University Press, Cambridge (1972).

D. H. Williams and I. Howe, *Principles of Organic Mass Spectrometry*, McGraw-Hill, London (1972).

SPECTRA CATALOGUES

A. Cornu and R. Massot, *Compilation of Mass Spectral Data*, Heyden, London, and Presses Universitaires de France, Paris (1966–8). (Lists the ten principal ions in the mass spectra of over 5000 compounds.)

J. H. Beynon and A. E. Williams, *Mass and Abundance Tables for Use in Mass Spectrometry*, Elsevier, Amsterdam (1963).

J. Lederburg, *Computation of Molecular Formulas for Mass Spectrometry*, Holden-Day, San Francisco (1964).

J. H. Beynon, R. A. Saunders and A. E. Williams, *Tables of Metastable Transitions for Use in Mass Spectrometry*, Elsevier, Amsterdam (1965). (These last three compilations form a complete kit for deducing molecular-ion and fragmention formulae from the mass spectrum m/e values.)

SUPPLEMENTARY TEXTS

G. Lawson and J. F. J. Todd. Radiofrequency Quadrupole Mass Spectrometers. *Chem. Brit.* (1971), 373.

G. R. Waller (ed.), *Biochemical Applications of Mass Spectrometry*, Wiley, Chichester (1972).

G. W. A. Milne (ed.), *Mass Spectrometry. Techniques and Applications*, Wiley, New York (1972).

6

SPECTROSCOPY PROBLEMS†

The first four sections of this chapter contain varied exercises in the application of each separate technique to an organic problem. The problems in the penultimate section (6.5) demand an interplay among these techniques. Solutions are given at the end of the chapter in section 6.6.

6.1 INFRARED SPECTROSCOPY PROBLEMS

(i) What are the assignments for the following absorption bands? (*Example*: figure 2.2, 2150 cm^{-1}. *Answer*: C≡C *str*)

Figure 2.3, 755 cm^{-1}, 3400 cm^{-1}. Figure 2.4, 1365 cm^{-1}, 3365 cm^{-1}. Figure 2.7, 1585 cm^{-1}, 1605 cm^{-1}. Figure 2.9, 700 cm^{-1}, 1720–1980 cm^{-1}, 2000–2600 cm^{-1}. Figure 2.14, doublet at 1600 cm^{-1}, doublet at 1390 cm^{-1}. Figure 2.15, 1625 cm^{-1}, 1660 cm^{-1}, 3175 cm^{-1}, 3375 cm^{-1}.

(ii) Identify on the spectrum listed the absorption band corresponding to the given vibration. (*Example*: figure 2.2, ≡C—H *str*. *Answer*: 3320 cm^{-1})

Figure 2.3, aromatic C—H *str*, aldehyde C—H *str*, C=O *str*. Figure 2.4, aromatic C⩵C *str*, C=O *str*, overtone (2*v*) of C=O *str*. Figure 2.6, free

---

† The n.m.r. spectra reproduced in this chapter are from *High Resolution N.M.R. Spectra Catalog*, with permission of the publishers, Varian Associates, Palo Alto, California, U.S.A.

The infrared spectra reproduced in this chapter were recorded on a Perkin–Elmer Model 700 infrared spectrophotometer.

O—H *str*, bonded O—H *str*. Figure 2.13, aliphatic C—H *def*, alkene out-of-plane C—H *def*. Figure 2.16, C=C *str*, N—H *def*, C—N *str*. Figure 2.17, out-of-plane C—H *def*, C≡N *str*. Figure 2.18, S=O *str*, both v_{anti} and v_{symm}.

(iii) Compound A has molecular formula C_3H_3N, and its infrared spectrum is given in figure 6.1. Suggest a structure for A.

(iv) Compound B has molecular formula $C_6H_{10}O$, and its infrared spectrum is given in figure 6.2. Suggest a structure for B.

(v) The liquid-film infrared spectrum of 2,4-pentanedione (acetylacetone) shows absorption bands at 1600 cm^{-1} (strong, broad); 1710 cm^{-1} (less strong than the 1600 cm^{-1} band, and more sharp); a very broad band stretching from $\approx 2400 \text{ cm}^{-1}$ to 3400 cm^{-1}, which is unchanged on dilution. What are these bands?

(vi) State whether the following pairs of compounds could be distinguished by an examination of their infrared spectra. Give reasons.

$PhCH_2NH_2$ and $PhCONH_2$
$H_2N—C_6H_4—CO_2Me$ and $Me—C_6H_4—CONH_2$
$MeO—C_6H_4—COMe$ and $Me—C_6H_4—CO_2Me$
cyclohexanone and 3-methylcyclopentanone
$PhCOCH_2CH_3$ and $PhCH_2COCH_3$

6.2 N.M.R. SPECTROSCOPY PROBLEMS

All the spectra in the following problems can be fully analysed as for first-order systems: chemical shift and coupling constant data should be extracted and compared with the values given in tables 3.4–3.10.

Worked solutions to the problems are provided in scrambled form on page 240. Work through each problem and use the stepwise assistance provided in the solutions only when necessary.

(i) The spectrum shown (figure 6.3) is that of the dental local anaesthetic procaine (at 60 MHz).

procaine

Assign all signals, accounting for (i) chemical shift values, (ii) integrals and (iii) coupling constants.

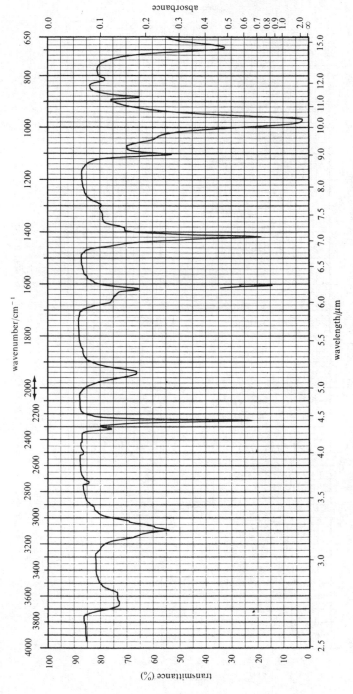

Figure 6.1 *Infrared spectrum for compound A in problem 6.1 (iii). Liquid film.*

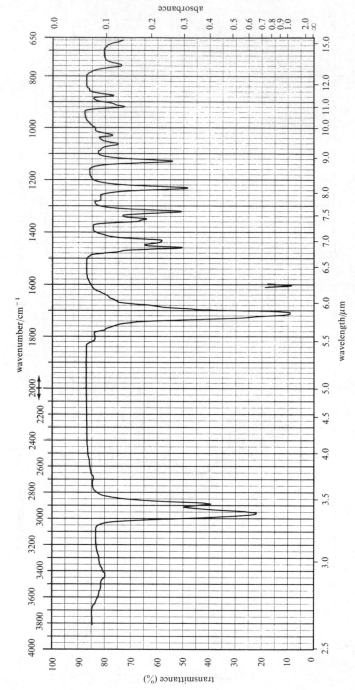

Figure 6.2 *Infrared spectrum for compound B in problem 6.1 (iv). Liquid film.*

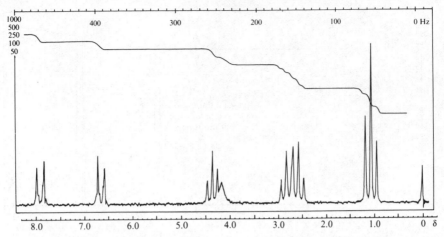

Figure 6.3 *N.M.R. spectrum for procaine in problem 6.2(i).*

(ii) The spectrum shown in figure 6.4 is that of 4-vinylpyridine (at 60 MHz).

N⟨pyridine ring⟩—CH=CH₂

4-vinylpyridine

Assign all signals, accounting for (i) chemical shift values, (ii) integrals and (iii) coupling constants.

(iii) The 60 MHz spectrum shown in figure 6.5 is that of a hydrocarbon C_9H_{12}. Deduce its structure by accounting for (i) chemical shift values, (ii) integrals and (iii) coupling constants.

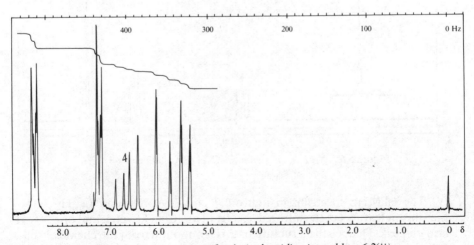

Figure 6.4 *N.M.R. spectrum for 4-vinylpyridine in problem 6.2(ii).*

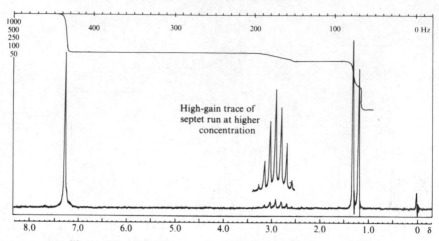

Figure 6.5 *N.M.R. spectrum for* C_9H_{12} *in problem 6.2(iii).*

(iv) The 60 MHz spectrum in figure 6.6 is of a compound $C_4H_7BrO_2$. Infrared evidence shows it to be a carboxylic acid. Deduce its structure by accounting for (i) chemical shift values, (ii) integrals and (iii) coupling constants.

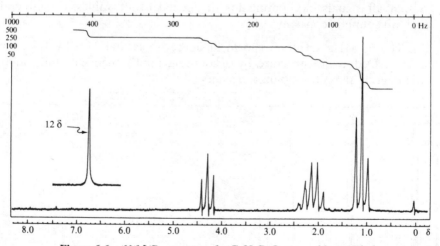

Figure 6.6 *N.M.R. spectrum for* $C_4H_7BrO_2$ *in problem 6.2(iv).*

(v) The 60 MHz spectrum in figure 6.7 is of a compound $C_{10}H_{13}NO_2$. Significant features of the infrared spectrum are C=O stretch and one N—H stretch peak. Deduce the structure of this compound from the chemical shift, integral and coupling data shown on the spectrum.

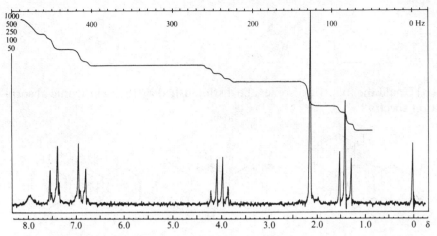

Figure 6.7 *N.M.R. spectrum for* $C_{10}H_{13}NO_2$ *in problem 6.2(v).*

6.3 ELECTRONIC SPECTROSCOPY PROBLEMS

(i) Use the Woodward rules in table 4.3 to predict the expected λ_{max} for the following compounds dissolved in ethanol.

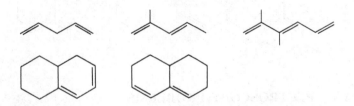

(ii) Use the Woodward rules in table 4.4 to predict the expected λ_{max} for the $\pi \rightarrow \pi^*$ transition in the following compounds (in ethanol).

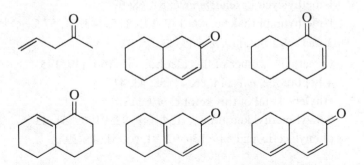

(iii) A ketone was known to have one of the isomeric structures shown overleaf and had λ_{max} (in ethanol) at 224 nm. Which was it?

(iv) Could the following isomers be distinguished by their electronic absorption spectra?

(v) For each of the following compounds, write out the structure of an isomer that is likely to have a substantially different electronic absorption spectrum.

6.4   MASS SPECTROSCOPY PROBLEMS

(i) Write feasible structures for these ions (found in the following mass spectra).

> 1-methylcyclohexene, m/e 96, 81, 68, 67
> 1-methylnaphthalene, m/e 142, 141, 115, 143, 71, 57.5
> 3-methyl-2-butanol, m/e 45, 43, 55, 73
> trimethylsilylether of this alcohol, m/e 160, 117, 145
> 4-heptanone, m/e 114, 86, 71, 58, 43, 41
> ethylene ketal of this ketone, m/e 113
> benzyl methyl ketone, m/e 134, 119, 92, 91, 65, 51, 43
> phenylacetic acid, m/e 136, 92, 91, 65, 51, 45, 39

(ii) Deduce feasible structures (not necessarily unambiguous) for the compounds whose mass spectra have ions at the following m/e values. Base peak first.

$C_{11}H_{16}$, m/e 91, 119, 148, 41, 27, 39, 92, 77, 51, 29
$C_{11}H_{14}O_2$ (ester), m/e 105, 123, 77, 56, 122, 106, 41, 29
$C_{10}H_{20}O$ (alcohol), m/e 57, 81, 67, 56, 82, 83, 41, 123, 99
$C_{10}H_{12}O_2$ (carboxylic acid), m/e 149, 164, 105, 119, 77, 91, 79, 131, 135, 150

Note : From the molecular formula, it is always useful to calculate the number of *double-bond equivalents* in the molecule. For an alkane, the molecular formula is C_nH_{2n+2}: for an alkene, or cycloalkane, it is C_nH_{2n}, so that loss of 2H (implying the presence of *either* one double bond *or* one ring) is one double-bond equivalent (D.B.E.). Dienes, cycloalkenes or bicyclo-alkanes have two D.B.E. Trienes etc., contain 3 D.B.E., benzene contains 4 D.B.E. To find the number of D.B.E. from molecular formula C_nH_mO, use the formula D.B.E. $= \frac{1}{2}[(2n + 2) - m]$. For formulae $C_nH_mN_p$, use D.B.E. $= \frac{1}{2}[(2n + 2) - (m - p)]$. Thus $C_{11}H_{16}$ contains 4 D.B.E. (probably an aromatic ring?).

(iii) Calculate the m/e value for the parent ions (m_1) that produce the following normal daughter ions (m_2) and metastable daughter ions (m^*).

$m_2 =$	117	$m^* =$	93.8
	61		31.8
	131		117.5
	77		56.5
	51		33.8

(iv) How could the following pairs of isomers be differentiated by their respective mass spectra?

and

and

and

6.5   CONJOINT I.R.–U.V./VIS.–N.M.R.–MASS SPECTROSCOPY PROBLEMS

General approach. Every spectroscopic problem of this type is unique, but you should begin by quickly perusing all of the spectral data and noting the following. Are there useful and prominent bands in the infrared spectrum (C=O, C≡C, C≡N, O—H, N—H, etc.)? Is the compound aromatic, alkene, alkane (n.m.r. and i.r. spectra)? Is there an ultraviolet chromophore, and can it be tentatively identified? Is there an $M^{\ddagger}$ peak in the mass spectrum, from which can be found the molecular formula and the number of D.B.E.?

Thereafter it will usually be found necessary to extract information from the spectra in succession, gradually homing in on an unequivocal solution. When a structure finally emerges, it should be examined scrupulously against all the available data until no shadow of doubt remains.

(i) When acetone is treated with base, a higher boiling liquid (b.p. 130°C) can be isolated from the reaction mixture. The spectroscopic properties of this liquid are: infrared, 1620 cm^{-1} (m), 1695 cm^{-1} (s): n.m.r., 1.9 δ (3H, singlet), 2.1 δ (6H, singlet), 6.15 δ (1H, singlet): u.v., λ_{max} 238 nm (ε_{max} 11 700): mass, m/e (R.A.), 55(100), 83(90), 43(78), 98(49), 29(46), 39(43), 27(42), 53(13), 41(13), 28(8).

Make a sketch of the n.m.r. spectrum and construct a mass spectrum bar diagram. Deduce the structure and account for all of the observed data.

(ii) Deduce the structure of the compound (m.p. 248°C) whose spectral data are given in figure 6.8. (λ_{max} typically that of a substituted benzene, around 250 nm, ε_{max} around 12 000.)

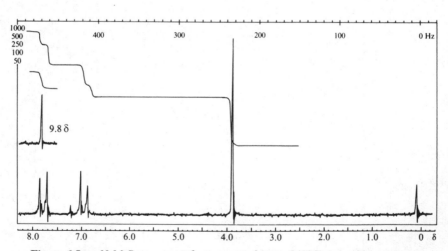

Figure 6.8a   *N.M.R. spectrum for compound* (m.p. 248°C) *in problem 6.5(ii).*

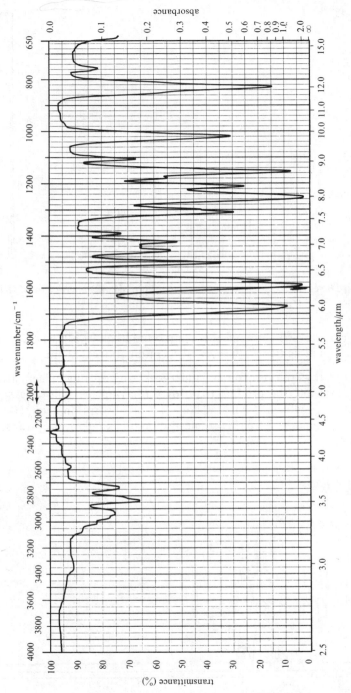

Figure 6.8b *I.R. spectrum for compound* (m.p. 248°C) *in problem 6.5(ii).*

238 ORGANIC SPECTROSCOPY

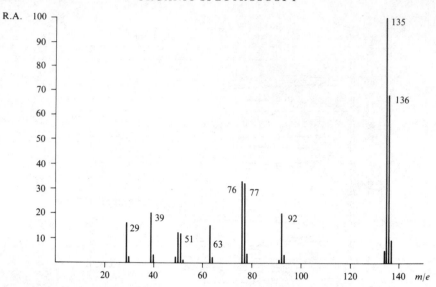

Figure 6.8c *Mass spectrum for compound (m.p. 248°C) in problem 6.5(ii).*

(iii) Deduce the structure of the compound (b.p. 97°C) whose spectral data are given in figure 6.9. (No u.v. absorption above 200 nm.)

(iv) 2,2-Dimethylcyclopropanone undergoes ring-opening when attacked by methoxide ion, the product (b.p. 101°C) having the following spectral properties: i.r., 1740 cm^{-1} (s), 1160 cm^{-1} (s), no absorption near 1600 cm^{-1} or 3100 cm^{-1}: n.m.r., 3.6 δ (3H singlet), 1.2 δ (9H singlet): u.v., transparent

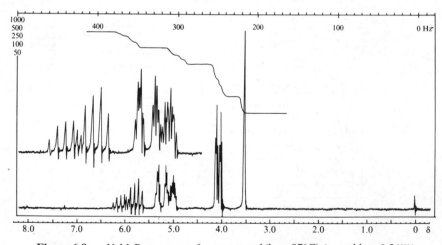

Figure 6.9a *N.M.R. spectrum for compound (b.p. 97°C) in problem 6.5(iii).*

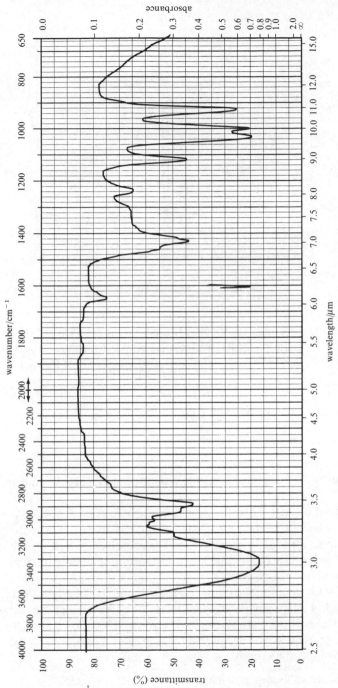

Figure 6.9b *I.R. spectrum for compound* (b.p. 97°C) *in problem 6.5(iii).*

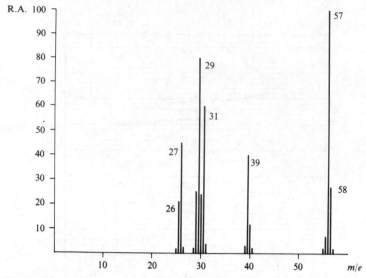

Figure 6.9c *Mass spectrum for compound* (b.p. 97°C) *in problem 6.5(iii).*

above 200 nm: mass, *m/e* 116, 85, 59, 31. Deduce the structure of the product and suggest a mechanism for its formation. From the mechanism, an alternative product might have arisen: what is it and why is it not formed?

6.6 SOLUTIONS TO PROBLEMS

6.1 Infrared spectroscopy problems

(i) and (ii) See correlation charts

(iii) acrylonitrile (C≡N *str*, C=C *str*, out-of-plane C—H *def* and its overtone at 2*v*).

(iv) cyclohexanone (C=O *str* for six-ring or acyclic ketone. Alkane C—H *str* and *def*. No alkene C—H *str*, or C=C *str*. No aldehyde C—H *str*.)

(v) 2,4-pentanedione is 85 per cent enolic; intramolecular hydrogen bonding in the enol gives O—H *str* and low C=O *str* (together with C=O *str*). The keto form has normal C=O *str*.

(vi) Yes (C=O *str* and N—H *def* (Amide I and II) in $PhCONH_2$; wide separation of N—H *str* bands in $PhCONH_2$). Yes (as before). No (or at least not with certainty since both show C=O *str* near 1700 cm^{-1}, and C—O *str*). Yes (C=O *str* frequency is dependent on ring size). Yes (C=O *str* frequency is dependent on conjugation).

6.2 N.M.R. spectroscopy problems

These solutions are scrambled to offer assistance to the student who is finding difficulty, but does not wish to see the final solution too soon. Identify the sequential steps in the solutions to each problem as i(1), i(2), i(3), etc.

i(1) The low-field peaks (above 6 δ) are aromatic; all others are aliphatic. Calculate the integrals. Deal with the aromatic peaks first, then the two ethyl groups, then the —CH_2CH_2— groups, then NH_2.

ii(1) Table 3.7 gives the δ values for all these protons. The integral ratios can only be significantly interpreted if you appreciate the multiplicity in the spectrum.

iii(1) The peak at 7.1 δ is aromatic H. The others are aliphatic. Are the two high-field peaks separate singlets, or do they together constitute a doublet?

iv(1) The presence of carboxyl is confirmed by the 12 δ peak. Taking COOH from the molecular formula leaves C_3H_6Br.

v(1) The compound is aromatic (peaks above 6 δ), and the broad singlet at 7.9 δ is probably H attached to nitrogen (infrared evidence confirms; see also table 3.8).

v(3) If 4H are on the ring and 1H is attached to N, this leaves 8H remaining.

iv(8) Compound is 2-bromobutyric acid (v).

iv(7) Predict the δ values for V. Predicted δ for CH_3 (table 3.4) is 0.9 δ; observed, 1.05 δ. Predicted δ for CH_2 (tables 3.4 and 3.5) is $(1.25 + 0.6)\,\delta = 1.85\,\delta$; observed, 2.1 δ. Predicted δ for CH (table 3.6) is $(1.25 + 0.7 + 1.9)\,\delta = 3.85\,\delta$; observed, 4.25 δ.

i(11) Predicted δ for protons c, CH_2—NR_2, is $(2.5 + 0.3)\,\delta = 2.8\,\delta$ (tables 3.4 and 3.5). The low-field 3 lines (centred on 2.8 δ, integral 2; J, 8 Hz) are the CH_2—N triplet; that is, this triplet overlaps with the ethyl quartet (same J value) giving an apparent quintet.

ii(6) The vinyl protons are coupling AMX, so that each proton gives rise to a quartet, integral 1. From the approximate chemical shift values in 5, we can make a preliminary allocation of the protons observed as

6.6 δ, quartet A   X 5.35 δ, quartet

M 5.85 δ, quartet

Although the predicted δ values show only moderate agreement with the observed values, the observed *relative* positions follow the predicted order. Final confirmation of the allocation rests with the AMX coupling analysis: measure the three J values, and note that each multiplet has two J values. Check these against the predicted values in table 3.10.

v(12) CH_3CONH is the second aromatic substituent.

iii(5) Predicted δ for CH_3 (tables 3.4 and 3.5) is $(0.9 + 0.3)\,\delta = 1.2\,\delta$. Coupling with methine (1H) gives a doublet at 1.2 δ, J, 8 Hz.

i(10) Using tables 3.4 and 3.5 the predicted δ for protons e, CH_2—OCOR, is $(4.15 + 0.1)\,\delta = 4.25\,\delta$. Observed triplet at $4.3\,\delta$, J, 8 Hz. Where is the CH_2—N triplet?

iv(4) Consider the predicted multiplicities and integrals for each of these.

ii(7) Predicted J_{AM}, 11–19 Hz; observed, 19 Hz. Predicted J_{AX}, 5–14 Hz; observed, 12 Hz. Predicted J_{MX}, 3–7 Hz; observed, ≈ 1.5 Hz.

iii(3) C_6 is accounted for in the benzene ring, leaving C_3 in *one* side-chain, which also contains 7H. Side-chain is C_3H_7.

iv(5) Only IV and V agree in multiplicity, but not in integrals, with the spectrum. (J, 8 Hz).

v(2) It is *p*-substituted (AA'BB' system, J_{ortho}, 11 Hz).

i(9) The —CH_2CH— group should give rise to two triplets, each integral 2; deal with e first, then c.

v(7) Fragments of the structure are

—NH   —CO— CH_3CH_2— CH_3—   [and —O—]

(One O atom has to be added to bring the total to the molecular formula).

i(8) The ethyl quartet is overlapped; the 4 lines at highest field, integral 4, contain the quartet. Thus for the two N-ethyl groups we see the triplet and quartet, J, 8 Hz. Predicted for CH_2, $2.5\,\delta$; observed, $2.6\,\delta$. Predicted for CH_3 (tables 3.4 and 3.5), $(0.9 + 0.1)\,\delta = 1.0\,\delta$; observed, $1.05\,\delta$.

v(11) CH_3CH_2O— is one aromatic substituent.

CH_3CO— and CH_3NHCO— cannot be aromatic substituents.

i(7) Table 3.4 gives $2.5\,\delta$ for CH_2—NR_2.

ii(2) The pyridine ring protons are predicted (table 3.7) at $8.5\,\delta$ (α to N), and $7.0\,\delta$ (β to N); the observed values are $8.5\,\delta$ (approximate doublet), and $7.2\,\delta$ (approximate doublet).

v(5) The singlet at $2.1\,\delta$ is methyl.

i(6) The two N-ethyl groups *must* give rise to a triplet and a quartet. The triplet (protons a, integral 6) is clearly at $1.05\,\delta$. Calculate where the quartets (protons b) are from table 3.4.

v(8) Which of these functions (or combinations of these functions), attached to CH_3, will cause the CH_3 to appear at $2.1\,\delta$? (See table 3.4.)

i(5) Predicted δ for protons f, from table 3.9 is $(7.27 - 0.8 + 0.15)\,\delta = 6.62\,\delta$: observed, $6.6\,\delta$.

v(10) Use table 3.9 to infer which are the likely substituents on the ring.

ii(4) The best analogy for the vinyl protons is to consider pyridine $\equiv$ Ph in table 3.7.

ii(3) The low-field 'doublets' (J, 9 Hz) are AA'BB' in type, and the coupling constant is appropriate for *ortho* protons. Integrals correspond to 2H for each 'doublet'.

iii(4) Side-chain must be isopropyl. Predicted δ for methine CH—Ph (table

3.4) is 2.87 δ. Coupling with 2Me (6H) gives septet at 2.87 δ, J, 8 Hz as in table 3.10.

iv(2) The C_3 residue can be *n*-propyl or isopropyl, and the Br can be attached to several points within these possibilities.

v(4) There is a clear ethyl system present (triplet at 1.3 δ, quartet at 4.0 δ; J, 8 Hz). This accounts for 5H, leaving 3H unaccounted.

i(3) The two low-field aromatic doublets are typical of *p*-substitution, being similar to AX, but more rigorously AA′BB′. From table 3.10, J_{ortho} is predicted as 10 Hz, which is also observed.

i(4) Predicted δ for proton g, from table 3.9 is $(7.27 + 0.8 − 0.15)$ δ $= 7.92$ δ. Observed, 7.85 δ.

v(6) Check all integrals; ratio is 1:2:2:2:3:3.

ii(5) The predicted values for the vinylic protons are similar to

With three different environments, what is the multiplicity of the vinylic system?

i(2) The integral ratios shown (from low field up) are: 1 (doublet): 1 (doublet): 1 (triplet): 1 (broad singlet): 3 (quintet?): 3 (triplet). Since this only totals 10H, and procaine has 20H, the integrals correspond to 2:2:2:2:6:6.

iv(6) Predict the δ values for CH_2Br and CH_2COOH in IV.

iii(2) The molecule is aromatic and clearly benzenoid. Integrals show that the aromatic peak corresponds to 5H, and therefore the ring is monosubstituted. This leaves 7H in the other peaks.

v(9) Use table 3.4 to decide which are the possible neighbours for the ethyl CH_2 (appearing at 4.0 δ).

i(12) The broad singlet at 4.15 δ is $ArNH_2$ (protons d); predicted from table 3.8, between 4.0 and 3.5 δ.

iii(6) Compound is isopropylbenzene (cumene).

iv(3) The possibilities are

v(13)   The compound is aceto-*p*-phenetidine (phenacetin).

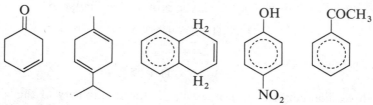

6.3   Electronic spectroscopy problems

(i)   < 200 nm, 225 nm, 255 nm, 235 nm, 273 nm.
(ii) < 200 nm, 227 nm, 227 nm, 242 nm, 338 nm, 286 nm.
(iii) The first structure.
(iv) The middle structure would have a biphenyl-like spectrum, while the other two would be indistinguishably naphthalene-like.
(v) Examples might be

6.4 Mass spectroscopy problems

(i) See cycloalkanes and cycloalkenes, sections 5.7.2 and 5.7.4.
See aromatic hydrocarbons (remembering tropylium ions and doubly charged ions), section 5.7.6.
See alcohols, section 5.7.8.
See section 5S.4. A_r for Si is 28.
See ketones, sections 5.7.11 and 5.7.13.
See section 5S.4.
See carboxylic acids, section 5.7.14.
(ii) 3-Phenylpentane.
Butyl benzoate.
4-*tert*-Butylcyclohexanol.
4-Isopropylbenzoic acid.
(iii) *m/e* 146, 117, 146, 105, 77
(iv) Prominent peak at M − C_3H_7.
Retro-Diels–Alder products.
Mass spectrum of ethylene ketals.

6.5   Conjoint spectroscopic problems

(i) $(Me)_2C=CHCOMe$ (mesityl oxide).
(ii) 4-Methoxybenzaldehyde (anisaldehyde).
(iii) $CH_2=CHCH_2OH$ (allyl alcohol).
(iv) Me_3CCO_2Me. Methoxide attacks the carbonyl group, with ring opening to give either a primary or a tertiary carbanion transition state: the latter is disfavoured so that $Me_2CHCH_2CO_2Me$ is not formed.

INDEX